ZEITSCHRIFT FÜR ANGEWANDTE MATHEMATIK UND MECHANIK

INGENIEURWISSENSCHAFTLICHE FORSCHUNGSARBEITEN

BAND 54 1974 HEFT 3

AUS DEM INHALT

HAUPTAUFSÄTZE

KLEINE MITTEILUNGEN

BUCHBESPRECHUNGEN

AKADEMIE-VERLAG GMBH · BERLIN

ZAMM Bd. 54 Nr. 3 S. 133—212 Berlin, März 1974

M 18,—

INHALT

Wir bitten, **Manuskriptsendungen** an folgende Anschrift zu richten:

Zeitschrift für Angewandte Mathematik und Mechanik
z. Hd. Herrn Prof. Dr. Günter Schmidt
Zentralinstitut für Mathematik und Mechanik
an der Akademie der Wissenschaften der DDR
DDR 108 Berlin, Mohrenstraße 39.

Zu den Arbeiten, die als Hauptaufsätze bestimmt sind, erbitten wir auf gesondertem Blatt eine kurze Zusammenfassung des Inhalts, nach Möglichkeit in deutscher, englischer und russischer Sprache; falls die Übersetzungen nicht geliefert werden können, ist wenigstens die Angabe spezieller Fachausdrücke in den verschiedenen Sprachen erwünscht. Ausführliche Hinweise für die Autoren, um deren strikte Berücksichtigung gebeten wird, finden sich im Anschluß an das Inhaltsverzeichnis des Jahrganges **53** (1973).

Zuschriften, welche Manuskripte und Arbeiten betreffen, deren Eingang von dem jetzigen Herausgeber Prof. Dr. H. Heinrich bestätigt worden ist, sind weiterhin an ihn unter der folgenden Anschrift zu richten:

Zeitschrift für Angewandte Mathematik und Mechanik
z. Hd. Herrn Prof. Dr.-Ing. Dr. h. c. Helmut Heinrich
DDR 8027 Dresden, Friedrich-Hegel-Str. 31.

Die Autoren erhalten von den Hauptaufsätzen 75, von den Kleinen Mitteilungen 25 Sonderdrucke ohne Berechnung, darüber hinaus bis zu 250 Sonderdrucke gegen Berechnung.

ZAMM 54, 133 – 138 (1974)

S. N. KATHURIA / G. L. KALRA

Effect of Tensor Conductivity on Rotating Superposed Plasmas

Die Stabilität einer Wirbelfläche zwischen zwei rotierenden hydromagnetischen Flüssigkeiten mit tensorieller Leitfähigkeit wird untersucht. Ebenso wird die Stabilität des Systems für den Fall diskutiert, daß es anfänglich ruht, aber einer zur Trennfläche senkrechten Beschleunigung unterliegt. Man findet, daß die tensorielle Leitfähigkeit das System destabilisiert, daß aber die Rotation je nach der Länge der Störungswellen stabilisierend oder destabilisierend wirken kann.

The stability of a vortex sheet between two hydromagnetic, rotating liquids having tensor conductivity is examined. The stability of the system without initial motion but subject to an acceleration perpendicular to the interface is also discussed. It is found that tensor conductivity destabilizes the system but rotation might be stabilizing or destabilizing depending upon the wave length of perturbation.

Исследуется стабильность вихревой поверхности между двумя вращающимися гидромагнитными жидкостями с тензорной проводимостью. Подвергается так же дискуссии стабильность системы для случая, если она, находясь первоначально в покое, подлежит перпендикулярному к поверхности раздела ускорению. Найдено, что тензорная проводимость приводит систему к неустойчивости и что вращение в зависимости от длины волн циркуляции может оказать как и стабилизирующее так и дестабилизирующее влияние.

1. Introduction

The linear stability of superposed hydromagnetic fluids has been a subject of extensive study [1]. The analysis of ideal hydromagnetic fluids shows that the magnetic field has a stabilizing influence on the hydrodynamic configuration [2].

JUKES [3] considered the effect of finite electrical resistivity on the stability of two fluids of different densities in a vertical gravitational field. He found that the inclusion of finite resistivity introduces unstable modes in a system stabilized by a uniform magnetic field.

MALLICK [4] considered the stability of a vortex sheet between two finitely conducting fluids for two-dimensional perturbations. He also found that inclusion of finite resistivity destroys the stabilizing effect of the magnetic field. CHAKRABORTY and JINDIA [5] extended the work of MALLICK to three-dimensional perturbations and showed that any instability which may be present for more general disturbances is also present for two-dimensional perturbations at a lower value of conductivity.

The above investigations adopt single fluid hydromagnetic plasma equations which, besides other assumptions, are valid for a characteristic frequency less than the ion cyclotron frequency. One can relax this assumption by modifying the equations to include finite ion cyclotron frequency. Several authors [6], [7], [8] incorporated the finite ion cyclotron frequency effects by modifying OHM's law so as to include tensor conductivity.

For high frequencies, tensor conductivity arises because the magnetic field is frozen with the electron-fluid rather than the ion-fluid. As in the case of scalar conductivity, it is found that tensor conductivity also gives rise to unstable modes [7], [8].

It is known that a uniform rotation has a stabilizing influence on the system of superposed hydromagnetic fluids [9]. It is, therefore, interesting to see whether presence of a uniform rotation can remove the instability due to tensor conductivity.

This paper considers a simple model where two inviscid, incompressible, homogeneous fluids with tensor conductivity form a vortex sheet. The magnetic field is assumed to be parallel to the vortex sheet. The entire configuration is subjected to a vertical (perpendicular to the interface) gravitational field and is rotating uniformly about an axis in the plane of the interface.

Our linear perturbation analysis shows that to first order the presence of tensor conductivity is destabilizing but rotation might have a stabilizing or destabilizing influence depending upon the wavelength of perturbations.

2. Basic Equations and the Dispersion Relation

Consider two incompressible, inviscid, homogeneous hydromagnetic fluids separated by a plane interface $z = 0$. The fluids occupy the semi-infinite regions $z \gtrless 0$ and are respectively denoted by the suffixes $+$, $-$. We assume the system to be subjected to an acceleration $\boldsymbol{g}$ $(0, 0, -g)$ which may account for the gravitational field or the curvature of the field lines. The whole configuration is rotating uniformly with angular velocity $\boldsymbol{\Omega}(0, \Omega, 0)$. Both the fluids carry a uniform impressed magnetic field $\boldsymbol{H}$ and have initial streaming velocities $\boldsymbol{U}_{\pm}$ in the xy-plane.

The equations for each fluid in a rotating frame of reference are [10]:

$$\varrho \frac{d\boldsymbol{u}}{dt} = -\nabla p + \frac{1}{4\pi}(\nabla\times\boldsymbol{H})\times\boldsymbol{H} + 2\,\varrho(\boldsymbol{u}\times\boldsymbol{\Omega}) + \varrho\,\boldsymbol{g}\,, \tag{1}$$

$$\nabla\cdot\boldsymbol{u} = 0\,, \qquad \boldsymbol{E} + \frac{1}{c}\boldsymbol{u}\times\boldsymbol{H} - \frac{1}{N e c}\boldsymbol{J}\times\boldsymbol{H} = 0\,. \qquad (2), (3)$$

The parameters $\varrho\,(= M\,N)$, N and p respectively denote the fluid density, particle number density and pressure which includes the centrifugal force term in the form of an equivalent effective pressure [10]. The vectors $\boldsymbol{u}$, $\boldsymbol{E}$, $\boldsymbol{J}$ represent the average flow velocity, electric field and current density respectively. The constants M, e and c denote ion mass, electronic charge and velocity of light respectively.

Eq. (3) is a modified OHM's law in which the last term accounts for the tensor conductivity and it arises due to the HALL drift [11]. It is assumed that interparticle collision frequency is small so that effects due to scalar conductivity may be ignored.

In addition to Eqs. (1)—(3), we have MAXWELL's equations,

$$\nabla\times\boldsymbol{E} = -\frac{1}{c}\frac{\partial \boldsymbol{H}}{\partial t}, \qquad \nabla\cdot\boldsymbol{H} = 0\,, \qquad \nabla\times\boldsymbol{H} = \frac{4\,\pi}{c}\boldsymbol{J}\,. \tag{4), (5), (6}$$

The steady state (suffix '0') of the configuration requires that

$$\frac{dp_0}{dz} = -\,g\,\varrho_0 + 2\,\varrho_0\,U_{0x}\,\Omega\,, \tag{7}$$

for each fluid and

$$p_0^+ = p_0^-\,, \tag{8}$$

at the interface.

In order to examine the stability of the initial state, consider two-dimensional perturbations

$$f \sim f_0(z)\,\exp\,(i\,k\,x + n\,t)\,, \tag{9}$$

where wave number, k is assumed to be real. A general study of three-dimensional perturbation makes the calculations very involved without substantially effecting the two-dimensional results; this is discussed in the appendix.

We linearize Eqs. (1)—(6) and write a differential equation governing h_z, the vertical component of the perturbed magnetic field.

$$(\psi^2 - k^2)\left[\psi^2 - k^2 - \left(\frac{\omega_c}{n'\,k\,a^2}\right)^2 (n'^2 + k^2\,a^2)^2\right] h_z = 0\,, \tag{10}$$

where $n'\,(= n + i\,\boldsymbol{k}\cdot\boldsymbol{U})$ is the DOPPLER shifted frequency, $a\left(= \frac{H_x}{\sqrt{4\,\pi\,\varrho}}\right)$ is the ALFVÉN speed, $\omega_c\ (= e\,H_x/Mc)$ is the ion cyclotron frequency and $\psi = \frac{d}{dz}$. Differential Eq. (10) is valid for both the fluids in the upper (+) and the lower (—) regions.

It may be remarked here that for perturbations perpendicular to the magnetic field, there is no effect of field since perpendicular perturbations simply exchange the lines of force while perturbation parallel to the field lines bend the lines of force and are important [2]. In view of this we take the direction of magnetic field parallel to the direction of perturbations. Again, the effect of initial streaming disappears for perpendicular perturbations and becomes pronounced for parallel propagation so that we take streaming also parallel to the direction of propagation.

The solutions of the differential equations (10), which are bounded at $z = +\,\infty$, for the upper (+) and $z = -\,\infty$ for the lower (—) fluid are,

$$h_{z+} = A_+\,e^{-kz} + B_+\,e^{-m_+z}\,, \qquad (z > 0)\,, \tag{11}$$

$$h_{z-} = A_-\,e^{kz} + B_-\,e^{m_-z}\,, \qquad (z < 0)\,, \tag{12}$$

where

$$m_\pm^2 = k^2\left[1 + \frac{(n_\pm'^2 + k^2\,a_\pm^2)^2}{(n_\pm'\,k^2\,a_\pm^2/\omega_c)^2}\right], \tag{13}$$

and it has been assumed without loss of generality that the real part of $m_\pm$ is positive.

To obtain the dispersion relation, we have to satisfy the appropriate boundary conditions at the surface of separation,

$$z = \varepsilon\,e^{ikx+nt}\,, \tag{14}$$

where ε is the displacement of the interface. The boundary conditions are obtained by integrating Eqs. (1)—(6). These are [12], [13] the continuity of (i) the fluid pressure, (ii) the normal component of the magnetic field, (iii) normal component of the fluid velocity and (iv) the tangential component of the magnetic field which is parallel to the impressed field. It can be seen that there are only four independent boundary conditions for the problem so that the four arbitrary constants $(A_\pm, B_\pm)$ can be eliminated to get the required dispersion relation,

$$\left[1+\frac{\varrho_+}{\varrho_-}\frac{n_+'^2}{n_-'^2}+\frac{g\,k}{n_-'^2}\left(1-\frac{\varrho_+}{\varrho_-}\right)-\frac{2\,i\,\Omega}{n_-'}\left(1-\frac{\varrho_+}{\varrho_-}\frac{n_+'}{n_-'}\right)\right]\left[1-\frac{n_+'^2}{n_-'^2}\frac{n_-'^2+a_-^2\,k^2}{n_+'^2+a_+'^2\,k^2}+\right.$$
$$+\frac{m_-}{k}\left\{1+\frac{m_+}{m_-}\frac{n_+'^2}{n_-'^2}\frac{n_-'^2+a_-^2\,k^2}{n_+'^2+a_+^2\,k^2}\right\}\Bigg]+\frac{2\,a_-^2\,k^2}{n_-'^2}\left[\frac{m_-}{k}\left\{1+\frac{m_+}{m_-}\frac{\varrho_+}{\varrho_-}\frac{n_+'^2}{n_-'^2}\frac{a_+^2}{a_-^2}\frac{n_-'^2+a_-^2\,k^2}{n_+'^2+a_+^2\,k^2}\right\}+\right.$$
$$\left.+\frac{g\,k}{n_-'^2}\left(1-\frac{\varrho_+}{\varrho_-}\right)-\frac{2\,i\,\Omega}{n_-'}\left(1-\frac{\varrho_+}{\varrho_-}\frac{n_+'}{n_-'}\frac{a_+^2}{a_-^2}\frac{n_-'^2+a_-^2\,k^2}{n_+'^2+a_+^2\,k^2}\right)\right]-$$
$$-\frac{2\,\varrho_+}{\varrho_-}\frac{n_+'^2}{n_-'^2}\left(1+\frac{2\,i\,\Omega}{n_+'}\right)\left(1-\frac{n_+'^2}{n_-'^2}\frac{n_-'^2+a_-^2\,k^2}{n_+'^2+a_+^2\,k^2}\right)=0\,. \tag{15}$$

Note that for $\Omega = 0$ Eq. (15) reduces to Eq. (28) of TALWAR and KALRA [8].

3. Discussion

It is convenient to discuss the dispersion relation for the two special cases, namely (1) KELVIN-HELMHOLTZ instability of an interface between identical fluids and (2) RAYLEIGH-TAYLOR instability of static fluids.

(1) KELVIN-HELMHOLTZ Instability

Consider two halves of a single homogeneous, incompressible plasma moving along x-axis with speeds U and $-U$ in the lower $(-)$ and upper $(+)$ regions respectively. For small HALL current ($\omega_c \to \infty$), Eq. (13) gives:

$$m_+ = \frac{[(n-i\,k\,U)^2 + a^2\,k^2]\,\omega_c}{k\,a^2\,(n-i\,k\,U)}\,, \qquad m_- = \frac{[(n+i\,k\,U)^2 + a^2\,k^2]\,\omega_c}{k\,a^2\,(n+i\,k\,U)}\,. \tag{16}$$

Using Eq. (16), the dispersion relation (15) in this case becomes

$$n\left[n^2 - k^2\left(U^2 - a^2 - \frac{2\,\Omega\,U}{k}\right)\right][\{n^2 - k^2\,(U^2 - a^2)\}^2 + 4\,k^2\,U^2\,n^2]$$
$$= \frac{2\,a^4\,U\,k^5}{\omega_c}[2\,n^2\,k\,U - \Omega\,(n^2 - k^2\,U^2 + k^2\,a^2)]\,. \tag{17}$$

For $\omega_c = \infty$, the dispersion relation (17) has the seven roots,

$$n = 0\,, \tag{18a}$$
$$n = \pm\,[k^2\,(U^2 - a^2) - 2\,\Omega\,k\,U]^{1/2}\,, \tag{18b, c}$$
$$n = i\,k\,(\pm\,a - U)\,, \tag{18d, e}$$
$$n = i\,k\,(\pm\,a + U)\,. \tag{18f, g}$$

The first root is just a trivial mode and the last four roots denote propagation of ALFVÉN waves at DOPPLER shifted frequencies. The rest of the two roots may lead to instability due to vortex sheet which can be stabilized by a magnetic field or rotation or both together. In order to see the effect of rotation in the presence of tensor conductivity, we calculate frequency shift due to large but finite ion cyclotron frequency. The frequency shifts in the respective roots of Eq. (18) are:

$$\Delta n = \frac{2\,a^4\,U\,\Omega\,k^2}{\omega_c\,(U^2 - a^2)}\,\frac{1}{k\,(a^2 - U^2) + 2\,\Omega\,U}\,, \tag{19a}$$
$$\Delta n = -\,a^4\,k^3/2\,\omega_c\,[k\,(a^2 - U^2) + 2\,\Omega\,U]\,, \tag{19b, c}$$
$$\Delta n = a^3\,k^2/4\,\omega_c\,(a \mp U)\,, \tag{19d, e}$$
$$\Delta n = a^3\,k^2/4\,\omega_c\,(a \pm U)\,. \tag{19f, g}$$

We find that to first order in ion cyclotron frequency, presence of a uniform rotation affects only the roots given by Eqs. (18a) and (18b, c). For sub-ALFVÉNic flow ($U/a < 1$) frequency shift of the root as given by Eq. (19a), does not introduce any instability. But for super-ALFVÉNic flow ($U/a > 1$), it introduces instability for wave-lengths $\lambda\left(=\frac{2\,\pi}{k}\right) > \lambda_{\text{crit.}}^{(1)}$ where

$$\lambda_{\text{crit.}}^{(1)} = \pi\,(U^2 - a^2)/\Omega\,U\,. \tag{20}$$

For a trans-ALFVÉNic flow ($U = a$), this mode is again found to be unstable and the growth rate is,

$$\Delta n = \frac{a^3\,k^3}{2\,\omega_c\,\Omega}\left(1 - \frac{\Omega}{2\,a\,k}\right). \tag{21}$$

It may be noted that this mode is stable for $\Omega \geqq 2\,a\,k$ and is unstable for $\Omega < 2\,a\,k$. The instability rate increases rapidly as $\Omega \to 0$.

Now consider the frequency shifts given by Eq. (19b, c). These shifts, which are identical, do not introduce any instability for a sub-ALFVÉNic flow ($U < a$). For a super-ALFVÉNic flow ($U > a$) also there is no instability if $\lambda > \lambda^{(1)}_{\text{crit.}}$ where $\lambda^{(1)}_{\text{crit.}}$ is defined by Eq. (20) but for $\lambda < \lambda^{(1)}_{\text{crit.}}$ the instability rate of the unstable mode is enhanced due to tensor conductivity, while the damping of the stable mode is reduced. For a trans-ALFVÉNic flow ($U = a$), these roots are stable.

Finally consider the frequency shifts given by Eq. (19d, e). The presence of a uniform rotation does not affect these modes but tensor conductivity introduces instability. For $a < U$, only one of the two modes is unstable but for $a > U$, both the modes are unstable. For $a = U$, the mode corresponding to Eq. (19d) disappears and that corresponding to Eq. (19e) is unstable. The root shifts given by Eq. (19f, g) also behave in the same way as discussed above.

(2) RAYLEIGH-TAYLOR Instability

To consider the stability of a density discontinuity between two static fluids subject to a uniform acceleration normal to the surface of separation, we rewrite Eq. (13) for $\omega_c \to \infty$,

$$m_\pm = \frac{n^2 + a_\pm^2 k^2}{n\,k\,a_\pm^2}\,\omega_c\,. \tag{22}$$

Using Eq. (22), the dispersion relation (15) becomes,

$$n\,(n^2 + a_-^2 k^2)\,(n^2 + a_+^2 k^2)\,[n^2 + 2\,\alpha_-\,a_-^2\,k^2 + (\alpha_- - \alpha_+)\,(g\,k - 2\,i\,\Omega\,n)] + \frac{a_-^2\,a_+^2\,k^4}{\omega_c}(\alpha_- - \alpha_+)\,[n^4\,(\alpha_- - \alpha_+) + (g\,k - 2\,i\,\Omega\,n)\,(n^2 + 2\,\alpha_-\,a_-^2\,k^2)] = 0\,, \tag{23}$$

where

$$\alpha_+ = \varrho_+/(\varrho_+ + \varrho_-)\,, \qquad \alpha_- = \varrho_-/(\varrho_+ + \varrho_-)\,. \tag{24}$$

Eq. (23) is a seventh degree polynomial and its roots for $\omega_c = \infty$, are

$$n = 0\,, \tag{25a}$$

$$n = \pm\,i\,a_-\,k\,, \tag{25b, c}$$

$$n = \pm\,i\,a_+\,k\,, \tag{25d, e}$$

$$n = i\,\Omega\,(\alpha_- - \alpha_+) \mp i\,\gamma^{1/2}\,, \tag{25f, g}$$

where

$$\gamma = \Omega^2\,(\alpha_- - \alpha_+)^2 + 2\,\alpha_-\,a_-^2\,k^2 + (\alpha_- - \alpha_+)\,g\,k\,. \tag{26}$$

In the presence of large but finite ω_c, the change in first root (Eq. 25a) is given by

$$\Delta n = -\frac{2\,a_-\,g\,k^3\,\alpha^-\,(\alpha_- - \alpha_+)}{\omega_c\,[g\,k\,(\alpha_- - \alpha_+) + 2\,\alpha_-\,a_-^2\,k^2]}\,. \tag{27}$$

There is no effect of rotation on this root but finite ion cyclotron frequency is destabilizing for $\alpha_- < \alpha_+$ if

$$\lambda < \lambda^{(2)}_{\text{crit.}}\,, \qquad \lambda^{(2)}_{\text{crit.}} = \frac{4\,\pi\,\alpha_-\,a_-^2}{g\,(\alpha_+ - \alpha_-)}\,. \tag{28}$$

The frequency shift of the four roots given by Eq. (25b, c, d and e) in the limit of large ω_c is same for all the roots. This is

$$\Delta n = \frac{\alpha_-\,a_-^2\,k^2}{2\,\omega_c}\,. \tag{29}$$

There is no effect of rotation but presence of finite ion cyclotron frequency destabilizes the system. The frequency change due to finite ion cyclotron frequency in the last two modes, given by Eq. (25f, g), is

$$\Delta n = -\frac{2\,\alpha_-\,\alpha_+\,a_-^2\,a_+^2\,k^4}{\omega_c\,[2\,\alpha_-\,a_-^2\,k^2 + (\alpha_- - \alpha_+)\,g\,k]_+}\left[1 \mp \frac{(\alpha_- - \alpha_+)\,\Omega}{\gamma^{1/2}}\right]. \tag{30}$$

A system which is initially ($\omega_c = \infty$, $\Omega = 0$) stable with $\alpha_- > \alpha_+$ remains stable in the presence of finite ion cyclotron frequency and rotation.

In the case when the system is potentially unstable ($\alpha_+ > \alpha_-$) with $\omega_c = \infty$ and $\Omega = 0$, the instability rate is enhanced due to finite ion-cyclotron frequency for $\lambda > \lambda^{(2)}_{\text{crit.}}$ where $\lambda^{(2)}_{\text{crit.}}$ is given by Eq. (28). For $\lambda < \lambda^{(2)}_{\text{crit.}}$, the system is stable for $\omega_c = \infty$ and $\Omega = 0$ and it continues to be stable when uniform rotation and finite ion-cyclotron frequency are included.

The simultaneous presence of rotation and finite ion-cyclotron frequency makes an unstable system in the absence of rotation over-stable for $\lambda > \lambda^{(3)}_{\text{crit.}}$ where

$$\lambda^{(3)}_{\text{crit.}} = \frac{8\,\pi\,\alpha_-\,a_-^2}{g\,(\alpha_+ - \alpha_-)}\,(1 - \beta)^{-1} \quad \text{with} \quad \beta = \left(1 - \frac{8\,\alpha_-\,a_-^2\,\Omega^2}{g^2}\right)^{1/2}. \tag{31}$$

For $\lambda = \lambda^{(3)}_{\text{crit.}}$, the system is stable for $\omega_c = \infty$ but the simultaneous presence of uniform rotation and finite ion-cyclotron frequency makes the system overstable and the real part of the growth rate increases as $\Omega^{-1/2}$. This is also true for $\lambda = \lambda^{(4)}_{\text{crit.}}$, where $\lambda^{(4)}_{\text{crit.}}$ is obtained replacing ε by $-\varepsilon$ in Eq. (31).

For $\lambda < \lambda^{(3)}_{\text{crit.}}$ but greater than $\lambda^{(2)}_{\text{crit.}}$ the presence of rotation makes the unstable system, overstable.

4. Conclusion

It is shown that the simultaneous presence of tensor conductivity due to HALL drift and a uniform rotation has a destabilizing influence on the stability of a vortex sheet between two identical hydromagnetic liquids. For a small rotation in the plane of the interface but transverse to the field lines, there appears an unstable mode growing as Ω^{-1} for wavelengths greater than a critical limit. For static superposed fluids which are potentially unstable, finite rotation can introduce overstability in the presence of ion-cyclotron frequency. For a system which is stable in the presence of rotation alone, there exists for $\omega_c \neq \infty$, a set of critical wavelengths for which the system is overstable. The real part of growth rate in these cases increases as $\Omega^{-1/2}$.

Acknowledgements

The authors are grateful to Professor F. C. AULUCK and Dr. P. K. DAS for their interest in the work. One of us (G.L.K.) acknowledges the support under the NOAA project No. E-2-70(N). The authors are also grateful to a referee whose comments lead to the appendix of the paper.

Appendix

We shall now briefly outline the effect of three-dimensional perturbations on the present problem. Let us consider the perturbations given by,

$$f \sim f_0(z) \exp(i\, \boldsymbol{k} \cdot \boldsymbol{x} + n\, t)\,, \tag{A-1}$$

where the wave vector $\boldsymbol{k}(k_x, k_y, 0)$ is assumed to be real. The differential equation Eq. (10) governing h_z is replaced by

$$(\psi^2 - k^2)^2 - \left[\frac{4\,\delta^2\,k_x^2\,\Omega^2}{n'^2 + a^2\,k_x^2} + \frac{\omega_c^2}{a^4\,k_x^2}\,(n'^2 + a^2\,k_x^2)\left(1 + \frac{S^2\,a^2\,k_x^2}{n'^2}\right)\right](\psi^2 - k^2) + 4\,\Omega^2\,\omega_c^2\,\delta^2/a^4 = 0\,, \tag{A-2}$$

where

$$k = (k_x^2 + k_y^2)^{1/2} \quad \text{and} \quad \delta = k_y/k_x, \quad S = 1 + \frac{2\,\delta\,\Omega\,a^2\,k_x^2}{\omega_c\,(n'^2 + a^2\,k_x^2)}\,. \tag{A-3}$$

The above equation holds good for both the fluids.

The solutions of the differential Eq. (A-2), which are bounded at $z = \pm\infty$ are:

$$h_{z+} = A_+\,\mathrm{e}^{-m_1 z} + B_+\,\mathrm{e}^{-m_2 z}\,, \qquad (z > 0)\,, \tag{A-4}$$

$$h_{z-} = A_-\,\mathrm{e}^{m_3 z} + B_-\,\mathrm{e}^{m_4 z}\,, \qquad (z < 0)\,, \tag{A-5}$$

where the real part of m_i $(i = 1 - 4)$ is assumed to be positive. The expressions for m_i $(i = 1 - 4)$ can be obtained from Eq. (A-2). The dispersion relation can be written making use of appropriate boundary conditions. This is

$$\beta_1\,\beta_4 = \beta_2\,\beta_3\,, \tag{A-6}$$

where

$$\begin{aligned}\beta_1 = {} & S_-\,k_x\,(m_1 + m_4) - \frac{\omega_c^2\,(n_+'^2 + a_+^2\,k_x^2)}{a_+^4\,k_x\,(m_3 + m_4)}\left\{S_+ - S_- - \frac{S_+\,a_-^4\,k_x^2\,(m_1^2 - k^2)}{\omega_c^2\,(n_-'^2 + a_-^2\,k_x^2)} + \frac{S_-\,a_+^4\,k_x^2\,(m_4^2 - k^2)}{\omega_c^2\,(n_+'^2 + a_+^2\,k_x^2)}\right\} \\ & + i\,\delta\,\omega_c\left(\frac{n_-'}{a_-^2} - \frac{n_+'}{a_+^2}\right)\left\{S_- - 1 + \frac{a_-^4\,k_x^2}{\omega_c^2}\,\frac{m_1^2 - k^2}{n_-'^2 + a_-^2\,k_x^2}\right\},\end{aligned} \tag{A-7}$$

$$\begin{aligned}\beta_2 = {} & S_-\,k_x\,(m_2 + m_4) - \frac{\omega_c^2\,(n_+'^2 + a_+^2\,k_x^2)}{a_+^4\,k_x\,(m_3 + m_4)}\left\{S_+ - S_- - \frac{S_+\,a_-^4\,k_x^2\,(m_2^2 - k^2)}{\omega_c^2\,(n_-'^2 + a_-^2\,k_x^2)} + \frac{S_-\,a_+^4\,k_x^2}{\omega_c^2}\,\frac{m_4^2 - k^2}{n_+'^2 + a_+^2\,k_x^2}\right\} + \\ & + i\,\delta\,\omega_c\left(\frac{n_-'}{a_-^2} - \frac{n_+'}{a_+^2}\right)\left\{S_- - 1 + \frac{a_-^4\,k_x^2}{\omega_c^2}\,\frac{m_2^2 - k^2}{n_-'^2 + a_-^2\,k_x^2}\right\},\end{aligned} \tag{A-8}$$

$$\begin{aligned}\beta_3 = {} & \varrho_+\,n_+'^2\left(-1 + \frac{a_+^4\,k_x^2}{\omega_c^2}\,\frac{m_3^2 + m_4^2 + m_3\,m_4 - k^2}{n_+'^2 + a_+^2\,k_x^2}\right)\frac{\omega_c^2\,(n_+'^2 + a_+^2\,k_x^2)}{a_+^4\,k_x^2\,(m_3 + m_4)} \times \\ & \times\left[S_+ - S_- - \frac{S_+\,a_-^4\,k_x^2}{\omega_c^2}\,\frac{m_1^2 - k^2}{n_-'^2 + a_-^2\,k_x^2} + \frac{S_-\,a_+^4\,k_x^2}{\omega_c^2}\,\frac{m_4^2 - k^2}{n_+'^2 + a_+^2\,k_x^2}\right] + S_+\,\varrho_-\,n_-'^2\,m_1\left(1 - \frac{a_-^4\,k_x^2}{\omega_c^2}\,\frac{m_1^2 - k^2}{n_-'^2 + a_-^2\,k_x^2}\right) + \\ & + S_-\,\varrho_+\,n_+'^2\,m_4\left(1 - \frac{a_+^4\,k_x^2}{\omega_c^2}\,\frac{m_4^2 - k^2}{n_+'^2 + a_+^2\,k_x^2}\right) + S_+\,\{2\,i\,\Omega\,k_x\,(\varrho_+\,n_+' - \varrho_-\,n_-') + (\varrho_- - \varrho_+)\,g\,k^2\} \times \\ & \times\left(1 - \frac{a_-^4\,k_x^2}{\omega_c^2}\,\frac{m_1^2 - k^2}{n_-'^2 + a_-^2\,k_x^2}\right) + \frac{S_+\,\delta\,\omega_c\,k_x\,H^2}{4\,\pi\,i}\left(\frac{n_-'}{a_-^2} - \frac{n_+'}{a_+^2}\right)\left(S_- - 1 + \frac{a_-^4\,k_x^2}{\omega_c^2}\,\frac{m_1^2 - k^2}{n_-'^2 + a_-^2\,k_x^2}\right),\end{aligned} \tag{A-9}$$

and

$$\begin{aligned}\beta_4 = {} & \varrho_+\,n_+'^2\left(-1 + \frac{a_+^4\,k_x^2}{\omega_c^2}\,\frac{m_3^2 + m_4^2 + m_3\,m_4 - k^2}{n_+'^2 + a_+^2\,k_x^2}\right)\frac{\omega_c\,(n_+'^2 + a_+^2\,k_x^2)}{a_+^4\,k_x^2\,(m_3 + m_4)} \times \\ & \times\left[S_+ - S_- - \frac{S_+\,a_-^4\,k_x^2}{\omega_c^2}\,\frac{m_2^2 - k^2}{n_+'^2 + a_-^2\,k_x^2} + \frac{S_-\,a_+^4\,k_x^2}{\omega_c^2}\,\frac{m_4^2 - k^2}{n_+'^2 + a_+^2\,k_x^2}\right] + S_+\,\varrho_-\,n_-'^2\,m_2\left(1 - \frac{a_-^4\,k_x^2}{\omega_c^2}\,\frac{m_2^2 - k^2}{n_-^2 + a_-^2\,k_x^2}\right) +\end{aligned}$$

$$+ S_- m_4 \varrho_+ n_+'^2 \left(1 - \frac{a_+^4 k_x^2}{\omega_c^2} \frac{m_4^2 - k^2}{n_+'^2 + a_+^2 k_x^2}\right) +$$

$$+ S_+ \{2 i \Omega k_x (\varrho_+ n_+' - \varrho_- n_-') + (\varrho_- - \varrho_+) g k^2\} \left(1 - \frac{a_-^4 k_x^2}{\omega_c^2} \frac{m_2^2 - k^2}{n_-'^2 + a_-^2 k_x^2}\right) +$$

$$+ \frac{S_+ \delta \omega_c k_x H^2}{4 \pi i} \left(\frac{n_-'}{a_-^2} - \frac{n_+'}{a_+^2}\right) \left(S_- - 1 + \frac{a_-^4 k_x^2}{\omega_c^2} \frac{m_2^2 - k^2}{n_-'^2 + a_-^2 k_x^2}\right). \tag{A-10}$$

The dispersion relation is quite involved. In order to estimate the effect of arbitrary propagation, we retain only first order terms in δ and $1/\omega_c$.

Under this approximation, the values of $m_{1,2,3,4}$ become

$$m_{1,3}^2 = k^2 , \tag{A-11}$$

$$m_{2,4}^2 = \frac{\omega_c^2 (n_\pm^2 + a_\pm^2 k_x^2)}{a_\mp^4 k_x^2 n_\mp^2} \left[1 + \frac{4 \delta \Omega a_\mp^4 k_x^4}{\omega_c (n_\mp^2 + a_\mp^2 k_x^2)^2} + a_\mp^4 k_x^4 n_\mp^2 / \omega_c^2 (n_\mp^2 + a_\mp^2 k_x^2)^2\right]. \tag{A-12}$$

For KELVIN-HELMHOLTZ instability the dispersion equation (17) for two-dimensional perturbations is replaced by,

$$(n - \delta k_x U) [n^2 - k_x^2 (U^2 - a^2 - 2 \Omega U|k_x)] [\{n^2 - k_x^2 (U^2 - a^2)\}^2 + 4 k_x^2 U^2 n^2] =$$
$$= (2 a^4 U k_x^5/\omega_c) [2 n^2 U k_x - \Omega (n^2 - k_x^2 U^2 + a^2 k_x^2)] , \tag{A-13}$$

for $\omega_c = \infty$, six roots of the dispersion relation (A-13) are identical with the roots (18b, c, d, e, f, g) and the remaining root is replaced by

$$n = k_y U . \tag{A-14}$$

It is clear from Eq. (A-14) that δ-term introduces instability and makes the marginal root unstable.

The presence of finite ion-cyclotron frequency and rotation modify this root to

$$n = k_y U + \frac{2 a^4 U \Omega k}{\omega_c (U^2 - a^2) [2 \Omega U + k_x (a^2 - U^2)]} . \tag{A-15}$$

For $k_y = 0$, Eq. (A-15) reduces to Eq. (19a).

We find that for sub-ALFVÉNic flow ($U/a < 1$), k_y introduces instability provided $k > k_y^*$ where

$$k_y^* = \frac{2 a^4 \Omega k_x^2}{\omega_c (a^2 - U^2)_+ [2 \Omega U + k_x (a^2 - U^2)]_+} . \tag{A-16}$$

But for super-ALFVÉNic flow ($U/a > 1$), the three-dimensional perturbation introduces instability for all wave-lengths.

For a trans-ALFVÉNic flow ($U = a$), this mode is again unstable and the growth rate,

$$n = a k_y + \frac{a^3 k_x^3}{2 \Omega w_c} \left(1 - \frac{\Omega}{2 a k_x}\right). \tag{A-17}$$

From Eq. (A-17) we find that k_y introduces instability in marginal mode for trans-ALFVÉNic flow for $\Omega = 2 a k_x$. However, for $\Omega > 2 a k_x$, the mode becomes unstable for

$$k_y > \frac{a^3 k_x^3}{2 \Omega \omega_c} \left(\frac{\Omega}{2 a k_x} - 1\right). \tag{A-18}$$

It may be mentioned that in the rest of the six roots Eqs. (18b–g) and in RAYLEIGH-TAYLOR configuration introduction of k_y does not alter earlier results.

References

1 GERWIN, R. A., Stability of the interface between two fluids in relative motion, Rev. Mod. Phys. **40**, p. 652 (1969).
2 CHANDRASEKHAR, S., Hydrodynamic and Hydromagnetic Stability (Oxford University Press, London, 1961, Chapters X and XI.
3 JUKES, J. D., On the Rayleigh-Taylor problem in magnetohydrodynamics with finite resistivity, J. Fluid Mech. **16**, p. 177 (1963).
4 MALLICK, D. D., Instability of a surface of discontinuity of velocity in a parallel uniform magnetic field, J. Fluid Mech. **16**, p. 187 (1963).
5 CHAKRABORTY, B. B. and JINDIA, R. K., Hydromagnetic stability of a plane vortex sheet in finitely conducting fluid, ZAMM **51**, p. 367 (1971).
6 TAYLER, R. J., The influence of Hall effect on a simple hydromagnetic stability problem, Nucl. Fusion (Supp) Pt. **3**, p. 877 (1962).
7 HOSKING, R. J., New Instabilities due to Hall effect, Phys. Rev. Lett. **15**, p. 344 (1965).
8 TALWAR, S. P. and KALRA, G. L., Combined Taylor and Helmholtz instability in hydromagnetic including Hall effect, J. Plasma Phys. **1**, p. 145 (1967).
9 KALRA, G. L. and TALWAR, S. P., Kelvin instability for rotating compressible hydromagnetic fluids, Mon. Not. R. Astr. Soc. **135**, p. 391 (1967).
10 LEHNERT, B., Magnetohydrodynamic waves under the action of the Coriolis force, Astrophys, J. **119**, p. 647 (1954).
11 TANG, J. Y. and SEEBASS, R., Effect of tensor conductivity on collisionless magnetogasdynamic flows, Phys. Fluids **12**, p. 381 (1961).
12 STIX, T. H., The theory of Plasma Waves, McGraw Hill Book Co. New York 1962.
13 HOSKING, R. J. and KALRA, G. L., Boundary conditions in the presence of Hall current or finite ion Larmor radius effects, J. Plasma Phys. **7**, p. 545 (1972).

Eingereicht am 14. 4. 1972 (rev. Fassung am 10. 5. 1973)

Anschrift: S. N. KATHURIA, Weather Radar Unit, India Meteorological Department, New Delhi-3, India
Dr. G. L. KALRA, Department of Physics and Astrophysics, University of Delhi, Delhi-7, India

ZAMM 54, 139 – 144 (1974)

A. R. ROBINSON / J. C. THOMPSON

Transient Stresses in an Elastic Half Space Resulting from the Frictionless Indentation of a Rigid Wedge-Shaped Die*)

*Für das im Titel genannte Problem erhält man eine im Rahmen der klassischen Elastizitätstheorie exakte Lösung, wenn man annimmt, daß der Halbraum homogen und isotrop ist und der Keil mit konstanter Geschwindigkeit eindringt. Für eine gegebene Form des Keiles und des elastischen Mediums bestimmt die Eindringgeschwindigkeit eindeutig die Geschwindigkeit der vom Rande weg sich ausdehnenden Kontaktzone. Diese Geschwindigkeit, verglichen mit den Geschwindigkeiten der dilatorischen, rotatorischen und der Oberflächenwellen im elastischen Medium legt fest, welcher der vier möglichen charakteristischen Übergangszustände der Spannungen sich entwickelt.
Jeder der vier Kontaktgeschwindigkeitsbereiche wird mit der Methode der selbst-ähnlichen Potentiale behandelt.*

*A solution which is exact within the framework of the classical theory of elasticity is obtained for the title problem assuming that the half space is homogeneous and isotropic, and that the die indents at a constant rate. If the shape of the die and the elastic medium are specified, the rate of indentation uniquely determines the outward speed of the edge of the expanding contact zone. It is the speed at which the contact zone expands, relative to the speeds of the dilational, rotational and surface waves in the elastic medium which determines which of four possible characteristic transient stress states will be developed.
Each of the four ranges of contact speed is treated by the method of self-similar potentials.*

Решение, являющееся точным в рамках классической теории упругости, получено для определённой заглавием этой статьи задачи, при предпосылке, что полупространство однородное и изотропное, и что чекан врезывается с постоянной скоростью. Если форма чекана и упругое средство специфицированы, то скорость врезывания однозначно определяет распространяющуюся от края контактную зону. Эта скорость по сравнению со скоростями расширяющихся, вращающихся и поворхностных волн в упругом средстве определяет, которое из четырёх возможных характеристических переходных стрессовых состояний станет развиваться.
Каждый из четырёх интервалов контактной скорости определяется методом само-подобных потенциалов.

1. Introduction

When two solids collide, the behavior of the media in the region of contact depends strongly on the speed of impact. For low speeds, a quasi-static, HERTZian analysis gives an accurate estimate of the magnitude and distribution of the contact stress [1]. However, for greater contact speeds, inertial forces should not be neglected. In this paper, a solution which takes full account of all transient effects is obtained for the frictionless indentation of a homogeneous isotropic, linearly elastic half space by a rigid wedge-shaped die. The results are exact within the limits of the classical theory of elasticity although the inevitable inelastic behavior of the half space in regions near mathematical singularities is not considered. Such singularities occur directly beneath the apex of an indenting wedge and, for certain indentation velocities, at the edge of the region of contact and on any MACH fronts beginning at those points [2].

In contact problems each surface point generally has a non-zero displacement at the instant it contacts the second body. KOSTROV [3] overlooked this in treating the wedge problem described above. Moreover, he incorrectly reasoned that the flow of power across the surface of the indenting die would be infinite if each edge of the expanding contact moved outward with a speed α which exceeded c, the speed of the RAYLEIGH (surface) wave in the half space. Fortunately, his solution for $\alpha < c$ is correct if the boundary conditions are properly applied. Moreover, the same method of solution may be applied conveniently if $\alpha > c$.

2. Method of Self-Similar Potentials

A rigorous development of the self-similar method is given elsewhere [2], [4]. Essentially what is demonstrated is that any linearly elastic wave propagation problem involving boundary tractions or displacement which are homogeneous functions of x, y and t can be solved by a pair of complex-valued potentials $\Phi(\theta_1)$ and $\Psi(\theta_2)$ which are functions of the θ_i defined by the equation[1])

$$t - \theta_i x - y \sqrt{c_i^{-2} - \theta_i^2} = 0 , \qquad i = 1, 2 , \tag{1}$$

where the c_i are defined later. Because Φ and Ψ satisfy the two-dimension wave equations identically, i.e.,

$$\left(\frac{\partial^2}{\partial x^2} + \frac{\partial^2}{\partial y^2} - \frac{1}{c_i^2}\frac{\partial^2}{\partial t^2}\right) \{\Phi, \Psi\} = 0 \tag{2}$$

*) This paper describes a portion of the results of the doctoral dissertation of the second author.

[1]) In this paper any radical of the form $\sqrt{f^2 - \theta^2}$ is considered to have a value of $+f$ at $\theta = 0$ and to be analytic in the entire θ plane except across cuts on the real θ axis outward from the points $\theta = \pm f$.

their real parts, φ and ψ must do likewise. This condition is both necessary and sufficient [4], [5], that the equations of motion be satisfied by the stresses determined from the Lamé equations

$$\sigma_{xx} = \lambda\left(\frac{\partial u_x}{\partial x} + \frac{\partial u_y}{\partial y}\right) + 2\,\mu\,\frac{\partial u_x}{\partial x}, \qquad \sigma_{xy} = \mu\left(\frac{\partial u_x}{\partial y} + \frac{\partial u_y}{\partial x}\right), \qquad \sigma_{yy} = \lambda\left(\frac{\partial u_x}{\partial x} + \frac{\partial u_y}{\partial y}\right) + 2\,\mu\,\frac{\partial u_y}{\partial y} \tag{3}$$

and the relations

$$\frac{\partial^n u_x}{\partial t^n} = \frac{\partial \varphi}{\partial x} + \frac{\partial \psi}{\partial y}, \qquad \frac{\partial^n u_y}{\partial t^n} = \frac{\partial \varphi}{\partial y} - \frac{\partial \psi}{\partial x}. \tag{4}$$

The value of n must be such that the left-hand side of each equation is a homogeneous function of degree -1. If, as is the case in wedge problem, u_y is a homogeneous function of degree $+1$, then n must be $+2$. In such a case, it follows from Eqs. (1), (3), and (4) that on the $y = 0$ surface

$$\sigma_{yy} = \operatorname{Re} \sigma_{yy}^*(\theta)\,, \qquad \sigma_{xy} = \operatorname{Re} \sigma_{xy}^*\,, \qquad v_y = \frac{\partial u_y}{\partial t} = \operatorname{Re} v_y^*(\theta) \tag{5}$$

where $\theta = \theta_1 = \theta_2 = t/x$, and σ_{yy}^*, σ_{xy}^* and v_y^* denote the complex valued functions whose derivatives with respect to θ are

$$\begin{aligned} \sigma_{xy}^{*\prime} &= \mu\,[2\,\theta\,\sqrt{a^{-2} - \theta^2}\,\Phi' + (b^{-2} - 2\,\theta^2)\,\Psi']\,,\\ \sigma_{yy}^{*\prime} &= \mu\,[(b^{-2} - 2\,\theta^2)\,\Phi' - 2\,\theta\,\sqrt{b^{-2} - \theta^2}\,\Psi']\,, \end{aligned} \tag{6}$$

$$v_y^{*\prime} = -\sqrt{a^{-2} - \theta^2}\,\Phi' + \theta\,\Psi'\,. \tag{7}$$

Here $a = c_1 = \sqrt{(\lambda + 2\,\mu)/\varrho}$ and $b = c_2 = \sqrt{\lambda/\varrho}$ where ϱ is the density and λ and μ the Lamé constants. For half-space problems, the absence of shear on the $y = 0$ surface requires that $\sigma_{xy}^* = \sigma_{xy}^{*\prime} = 0$. In such cases, $v_y^{*\prime}$ can be expressed directly in terms of $\sigma_{yy}^{*\prime}$ as

$$v_y^{*\prime} = -\sqrt{a^{-2} - \theta^2}\,\sigma_{yy}^{*\prime}/(\mu\,b^2\,R(\theta^2)) \tag{8}$$

where $R(\theta^2)$ denotes the Rayleigh function

$$R(\theta^2) = (b^{-2} - 2\,\theta^2)^2 + 4\,\theta^2\,\sqrt{a^{-2} - \theta^2}\,\sqrt{b^{-2} - \theta^2}\,. \tag{9}$$

The properties of $R(\theta^2)$ will be considered later.

The essential consequence of the preceding discussion is that all half-space problems involving normal tractions and surface velocities which are self-similar (i.e., homogeneous of degree zero in x, y and t) may be solved by determining functions σ_{yy}^* and v_y^* whose real parts satisfy the boundary conditions and whose derivatives satisfy Eq. (8).

3. Boundary and Initial Conditions

The wedge apex angle γ is assumed to be shallow enough that the usual assumptions of the small displacement theory of elasticity are satisfied and also that displacements and stresses normal to the face of the wedge are essentially normal to the surface of the half space. Thus the absence of friction can be considered equivalent to the absence of shear on the surface. From [3], it can be anticipated that each edge of the expanding contact zone will move outward with a speed α which remains constant as long as V, the indentation velocity of the die, remains constant. Thus if contact begins at time $t = 0$ and at $x = y = 0$

$$\begin{aligned} &u_x \equiv u_y \equiv 0 \qquad \text{for all} \quad t < 0 \qquad \text{and for} \quad t > 0\,,\\ &\sigma_{xy}(x, 0, t) = 0\,, \quad 0 \leqq |x| \leqq \infty; \qquad \sigma_{yy}(x, 0, t) = 0\,, \quad \alpha\,t \leqq |x|\,,\\ &u_y(x, 0, t) = \begin{cases} V\,t - |x| \tan\dfrac{\pi - \gamma}{2} & |x| \leqq \alpha\,t\,,\\ 0\,, & |x| > a\,t \quad \text{if} \quad \alpha < a\,,\\ 0\,, & |x| > \alpha\,t \quad \text{if} \quad a < \alpha\,. \end{cases} \end{aligned} \tag{10}$$

The dependence of α on V and γ remains to be determined.

4. Additional Conditions of the Problem

In addition to causing Eqs. (10) to be satisfied, the normal traction must be such that

1) the deformed position of the $y = 0$ surface lies beneath the die at every point beyond the contact zone,
2) the stress field in the half space contains no singularity associated with a self-equilibrating disturbance, and
3) the force pressing the die into the half space is finite.

5. The Coupled Integral Equations

Taking into account the conclusions of Sec. 2 and the fact that the boundary conditions defined by Eqs. (10) are homogeneous functions of x and t, permits Eqs. (10) to be rewritten as

$$\sigma_{xy} \equiv \sigma^*_{xy} \equiv \sigma^{*\prime}_{xy} \equiv 0\,, \tag{11.a}$$

$$\sigma_{yy}(x, 0, t) = \operatorname{Re} \int_0^\theta \sigma^{*\prime}_{yy}\, d\theta = 0\,, \qquad |\theta| \leqq \alpha^{-1}\,, \tag{11.b}$$

$$u_y(x, 0, t) = \operatorname{Re} \int_0^t \int_0^\theta v^{*\prime}_y\, d\theta\, dt = V\, t - |x| \tan\frac{\pi - \gamma}{2}\,, \quad |\theta| > \alpha^{-1}\,, \tag{11.c}$$

The solution of these coupled integral equations is made more tractable by differentiating Eq. (11.c) to obtain expressions for the velocity and slope[2]).

$$v_y(x, 0, t) = \operatorname{Re} \int_0^\theta v^{*\prime}_y\, d\theta = V\,, \qquad |\theta| < \alpha^{-1}\,, \tag{12.a}$$

$$\frac{\partial u_y}{\partial x}(x, 0, t) = \frac{1}{2} \operatorname{Re} \int \theta\, v^*_y\, d\theta = \tan\frac{\pi - \gamma}{2}\,, \qquad \theta > \alpha^{-1}\,. \tag{12.b}$$

Since $\sigma^{*\prime}_{yy}$ and $v^{*\prime}_y$ can be non-analytic only at points on the real $\theta = t/x$ axis where $\partial\sigma_{yy}/\partial t \neq 0$ and $\partial v_y/\partial t \neq 0$

a) $\sigma^{*\prime}_{yy}$ must be analytic for all $|\theta| < \alpha^{-1}$,

b) $v^{*\prime}_y$ and $\sqrt{a^{-2} - \theta^{2\prime}}\, \sigma^{*\prime}_{yy}/R(\theta^2)$ must be analytic for all $|\theta| > \alpha^{-1}$ and

c) $v^{*\prime}_y$ must be analytic for $|\theta| < \alpha^{-1}$ if $\alpha < a$ (subseismic contact), or for $|\theta| < \alpha^{-1}$ if $\alpha > a$ (superseismic contact).

We now show that these conditions, together with those of Sect. 3, uniquely determine $\sigma^{*\prime}_{yy}$ and $v^{*\prime}_y$ for any value of α between zero and infinity. For convenience, the analysis considers separately the four ranges α bounded by the values 0, c, b, a and ∞ since $R(\theta^2)^{-1}$ is nonanalytic for $a^{-1} \leqq |\theta| \leqq b^{-1}$, and at $\theta = \pm c^{-1}$, where it has simple poles corresponding to the simple zeros in $R(\theta^2)$. These are shown in Fig. 1.

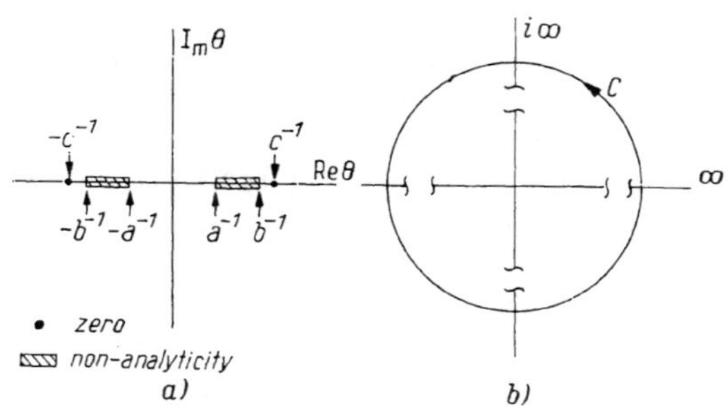

Fig. 1. Surface deformation and wave front positions for $\alpha < c$

6. The Lowest Range of Contact Speeds

If the die indents slowly enough that $\alpha < c$, the range of analyticity of $v^{*\prime}_y$ will not include any point where $R(\theta^2)^{-1}$ is non-analytic. Consequently, the conditions listed above will be satisfied if $\sigma^{*\prime}_{yy}$ and $\sqrt{a^{-2} - \theta^2}\, \sigma^{*\prime}_{yy}$ are analytic in the ranges $|\theta| < a^{-1}$ and $|\theta| > \alpha^{-1}$, respectively. These restrictions plus the symmetry of the problem require only that $\sigma^{*\prime}_{yy}$ be a function of the form

$$\sigma^{*\prime}_{yy} = \sum_{-\infty}^{+\infty} i\, F_n(\theta^2)\, (a^{-2} - \theta^2)^{-n-1/2} \tag{13}$$

where $i^2 = -1$, n is an integer, and each $F_n(\theta^2)$ is an integral (entire) function.

The form of $\sigma^{*\prime}_{yy}$, and hence, of each F_n is further constrained by the requirement that $\sigma^{*\prime}_{yy}$ be bounded at infinity, and elsewhere have no singularity exceeding a pole of order two. This eliminates the possibility of a stress field containing any self-equilibrating singularity [2]. It follows then that $\sigma^{*\prime}_{yy}$, and hence σ_{yy}, must be of the form

$$\sigma^{*\prime}_{yy}(\theta) = i\, A_1\, (\alpha^{-2} - \theta^2)^{-1/2} + i\, A_2\, (\alpha^{-2} - \theta^2)^{-3/2}\,, \tag{14}$$

$$\sigma_{yy}(x, 0, t) = -A \cosh^{-1}\frac{\alpha\, t}{|x|} - A_2\, \frac{\alpha^3\, t}{\sqrt{\alpha^2\, t^2 - x^2}}\,, \tag{15}$$

where A_1 and A_2 must be real-valued, positive constants for the contact stress to be compressive at every point. Their dependence on the γ and v follows from Eqs. (8) and (12.b). We find that

$$A_1 = 4\, \mu\, \pi^{-1} \left(1 - \frac{b^2}{a^2}\right) \tan\frac{\pi - \gamma}{2} \tag{16}$$

whatever the value of A_2. This indicates that the traction associated with A_1 causes a wedge-shaped displacement (Fig. 2) whereas the traction associated with A_2 causes an expanding punch-like displacement

[2]) In [3], the relationship $\alpha = V \tan(\pi - \gamma)/2$ was assumed equivalent to Eq. (12.b).

which has no cusp at the origin (Fig. 3). Obviously any superposition of these displacements violates the requirement that the deformed shape of the $y = 0$ surface lies below the die beyond the region of contact. Hence, A_2 must be identically equal to zero. As a result, the contact stress for all $\alpha < c$ is given by the equation

$$\sigma_{yy}(x, 0, t) = \begin{cases} -A_1 \cosh^{-1}\dfrac{\alpha t}{|x|}, & |x| < \alpha t, \\ 0, & |x| > \alpha t. \end{cases} \tag{17}$$

The stresses and velocities at any point in the half space may now be computed from Eqs. (1), (3), (4), (6), and (14). In essence, then, the problem of a slowly indenting wedge ($\alpha < c$) is completely solved by the function $\sigma^{*\prime}_{yy}$ defined by Eq. (14).

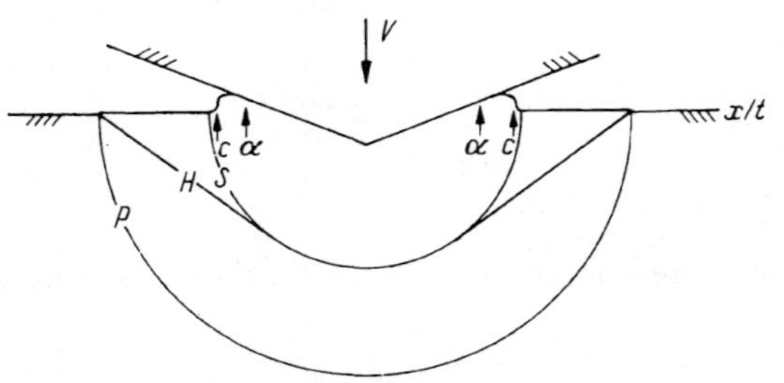

Fig. 2. a) Zero points, non-analyticities of $R(\theta^2)$
b) Integration contour of Eq. (14.b)

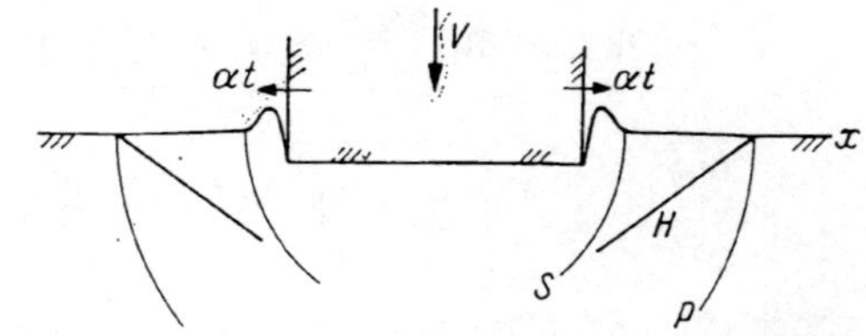

Fig. 3. Punch-like displacement for $\sigma_{yy} = -A_2 \alpha^3 t/\sqrt{\alpha^2 t^2 - x^2}$

7. Solution for the Remaining Subseismic Speeds $a > \alpha > c$

If the indentation velocity of the wedge is such that $\alpha > c$, then Eq. (14) is no longer an adequate definition of $\sigma^{*\prime}_{yy}$ since it leads to non-analyticities in $v^{*\prime}_y$ owing to the zeros and non-analyticities of $R(\theta^2)$ which occurs in the denominator of Eq. (8). The form of that equation and the restrictions on the range of non-analyticities of $\sigma^{*\prime}_{yy}$ and $v^{*\prime}_y$ suggest that a solution for speed $a > \alpha > c$ will be of the form

$$\sigma^{*\prime}_{yy}(\theta) = i\, A_1\, G(\theta)/\sqrt{a^{-2} - \theta^2} \tag{18}$$

provided that a correction factor $G(\theta^2)$ exists such that

(i) $G(\theta^2)$ is analytic except for $|\theta| > \alpha^{-1}$,
(ii) $G(\theta^2)/R(\theta^2)$ is analytic for $|\theta| < a^{-1}$ and $|\theta| > \alpha^{-1}$, and
(iii) $\lim\limits_{\theta \to \infty} G(\theta^2)/R(\theta^2) = 1$.

It is not difficult to deduce that $G(\theta^2)$ is uniquely defined by

$$G(\theta^2) = \begin{cases} (c^{-2} - \theta^2)\,(\alpha^{-2} - \theta^2)^{-1}, & b \geqq \alpha \geqq c, \\ (c^{-2} - \theta^2)\, F(\theta^2)\,(\alpha^{-2} - \theta^2)^{-1}, & a \geqq \alpha \geqq b, \end{cases} \tag{19}$$

where

$$F(\theta^2) = \exp\left(\frac{1}{\pi}\int\limits_{-\alpha^{-2}}^{b^{-2}} \frac{\operatorname{Im}\,[\log R(\lambda^2)]}{\lambda^2 - \theta^2}\, d\lambda^2\right) \tag{20}$$

For $b > \alpha > c$, Eqs. (18) and (19) lead to a contact stress

$$\sigma_{yy}(x, 0, t) = -A_1\left(\cosh\frac{\alpha t}{|x|} + \frac{\alpha t\,(\alpha^2 - c^2)}{c^2\sqrt{\alpha^2 t^2 - x^2}}\right). \tag{21}$$

For $a < \alpha < b$, the stress is most conveniently found by numerical integration [2]. The distribution of the contact stress shown in Fig. 4 has been computed in this way.

8. Solution for Superseismic Speeds ($\alpha > a$)

If the contact is superseismic, i.e., if $\alpha > a$, no disturbance propagates along the boundary more rapidly than the edge of the region of contact. For such problems

$$v_y(x, 0, t) = \begin{cases} 0, & |x| > \alpha t, \\ V, & |x| < \alpha t. \end{cases} \tag{22}$$

Since v_y is constant except at the points $x = \pm\,\alpha t$ where there are step-discontinuities of magnitude V, $v^{*\prime}_y$ must have simple poles at $\theta = \pm\,\alpha^{-1}$ and be analytic and imaginary-values at all other points on the real θ axis. These conditions together with Eqs. (8) and (12.b) require that superseismic contact problems be solved by

$$v_y^{*\prime} = \frac{-V\alpha}{i\pi(1-\alpha\theta)} + \frac{V\alpha}{i\pi(1+\alpha\theta)} \tag{23}$$

and

$$\sigma_{yy}^{*\prime} = \frac{-\mu b^2}{i\pi} \frac{R(\theta^2)}{\sqrt{\alpha^{-2}-\theta^2}} \frac{2\alpha^{-1} V}{\alpha^{-2}-\theta^2}. \tag{24}$$

The lengthy expressions which define $\sigma_{yy}(x, 0, t)$ are given in Ref. [2].

The simple poles in Eq. (24) indicate that each surface point experiences a step discontinuity in σ_{yy} at the instant at which it comes in contact with the die. The magnitude of the normal stress at this point then remains constant until the arrival of the front of the P wave. Thereafter, the stress varies continuously in the manner shown in Fig. 4.

The four cases considered above completely solve the problem of a rigid wedge indenting a linearly elastic, homogeneous half space, since the stress and velocity fields can be computed by a simple quadrature using Eqs. (1), (3), (4), (6) and the equation which defines $\sigma_{yy}^{\prime *}$ for the problem in question.

This approach was used in [2] to carry out a detailed analysis of the character of the stress and velocity fields near the edges of the contact zone, beneath the apex of the die, and immediately behind the various wave fronts. It is particularily noteworthy that singularities in the stress field beneath the surface occur only for problems in which $b < \alpha < a$ and then only along the Mach surface shown in Fig. 4. The order of the singularity depends on α and varies between 0 and $-1/2$.

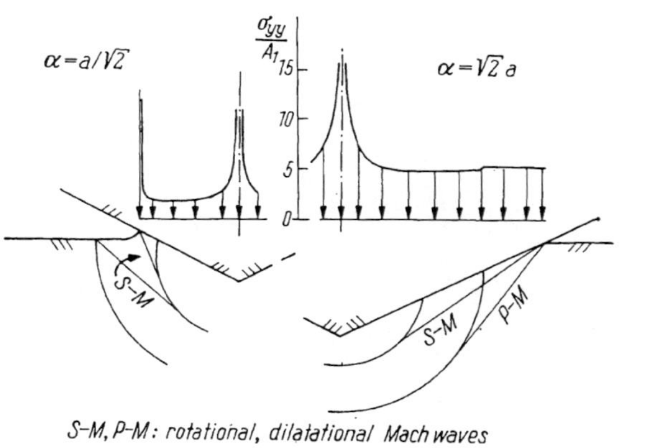

Fig. 4. Typical contact stress distributions and surface deformations ($\nu = 0.285$)

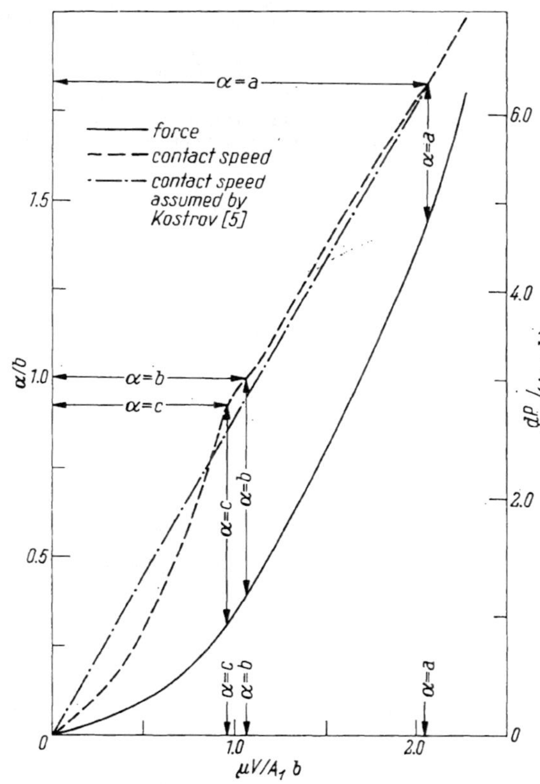

Fig. 5. Dependence of indentation force and contact speed on indentation velocity

9. Dependence of the Contact Speed on the Indentation Velocity

Because $\sigma_{yy}^{*\prime}$ is a known function of α, and $v_y^{*\prime}$ and $\sigma_{yy}^{*\prime}$ are related by Eq. (8), the dependence of α on V can be established simply by evaluating the integral of Eq. (8). The result obtained by a numerical integration procedure which assumes $b^2/a^2 = 0.3$ is shown in non-dimensional form in Fig. 5.

10. The Force Indenting the Wedge

The force required to press the wedge into the half space in any range of contact speed can be computed from the following formula which was derived in [2]

$$P(t) = i\pi t\, \sigma_{yy}^{*\prime}(\theta)|_\theta = 0\,. \tag{25}$$

A non-dimensional plot of $\partial P/\partial t$ vs. V is shown in Fig. 5, the calculation again being made for $b^2/a^2 = 0.3$.

11. Conclusions

Depending on the rate at which a rigid wedge-shaped die is pressed into the linearly elastic, isotropic half-space, one of four possible characteristic transient stress states will be developed. Each solution is applicable to a different range of outward speeds of the expanding contact region. The bounds of these four non-overlapping regions are zero, the speeds of the surface, rotational and dilational waves in the half-space, and infinity.

Examples are presented of the distribution of the contact stress in each range, and of the dependence of the indentation force and contact speeds on the indentation velocity.

References

1 LOVE, A. E. H., A Treatise on the Mathematical Theory of Elasticity, Fourth Edition. Dover, New York, 1944, p. 193.
2 THOMPSON, J. C. and ROBINSON, A. R., Exact Solutions of Some Dynamic Problems of Indentation and Transient Loadings of an Elastic Half-Space, Technical Report, Structural Research Series NO. SRS-350, University of Illinois, Urbana), Illinois 1969.
3 KOSTROV, B. V., Self-similar Dynamic Problems of Pressing of a Rigid Die into an Elastic Half Space, (in Russian). Mekhanika I Mashinostroenie 4, p. 54 (1964).
4 FRANK, PH. and VON MISES, R., Differential and Integral Equations of Mathematical Physics, Chapter XII, by S. L. Sobolev (Russian edition only), ONTI, Moscow/Leningrad 1937.
5 SMIRNOV, V. I. and SOBOLEV, S. L., On the Application of a New Method of Investigation of the Elastic Vibrations in the Space with Axial Symmetry, Trud. Inst. Seism. Akad. Nauk SSR 29 (1933).

Eingereicht am 13. 11. 1972, gekürzte Fassung: 17. 5. 1973

Anschrift: Professor A. R. ROBINSON, Civil Engineering Department, University of Illinois, Urbana, Illinois, U.S.A.
Professor J. C. THOMPSON, Civil Engineering Department, University of Waterloo, Waterloo, Ontario, Canada

ZAMM 54, 145–155 (1974)

B. NOVOTNÝ

Geometrically Nonlinear Shell Theory in the Light of the Method of Asymptotic Integration

Die in der Arbeit beschriebene Methode der asymptotischen Integration gestattet ein iteratives Vorgehen zur sukzessiven Bestimmung einer Näherungslösung der dreidimensionalen Gleichungen der geometrisch nichtlinearen Schalentheorie. Es wird eine Klassifikation der Schalenprobleme in Ausdrücken qualitativer und quantitativer Charakteristiken der Schalendaten dargeboten und auf eine Möglichkeit hingewiesen, den Einfluß verschiedener Typen von äußeren Einwirkungen abzuschätzen. Auf der durch die asymptotische Analysis gewonnenen Grundlage kann eine allgemeine Information über das Verhalten der Schale abgeleitet werden. Für gewisse Beispiele von Platten und Schalen werden die auf diesem Wege gemachten Voraussetzungen mit numerischen Ergebnissen verglichen.

The method of asymptotic integration, described in the paper, allows an iteration procedure for the successive determination of an approximate solution of threedimensional equations of geometrically nonlinear theory for the shell continuum. The classification of the shell problems is presented in terms of qualitative and quantitative characteristics of the shell data and a possibility of estimating the influence of different types of external actions is indicated. On the basis of the results yielded by the asymptotic analysis, a general information on the shell behaviour may be deduced. A comparison of the predictions made in this way with the results of the numerical analysis is given for some plate and shell examples.

Описанный в работе метод асимптотического интегрирования позволяет итерационным образом последовательно определить приближённое решение трёхмерных уравнений геометрически нелинейной теории оболочек. Предлагается классификация расчётных методов оболочек на основе выражений качественных и количественных характеристик параметров оболочек и приведена возможность оценки влияния разного рода внешних воздействий. Пользуясь результатами асиммптотического анализа возможно получить общую информацию о поведении оболочки. Произведено сравнение полученных этим путём выводов с численными результатами для определённых примеров плит и оболочек.

1. Introduction

The primary purpose of this paper[1]) is to describe and to discuss the way in which the method of asymptotic integration may be applied to furnish new information on the conditions limiting the use of the classical two-dimensional shell theories as theories providing the lowest order asymptotic approximation to the three-dimensional solution of the geometrically nonlinear theory of the elastic shell continuum.

The basic ideas and various modifications of the method of asymptotic integration are well known from the extensive literature on the application of this method to approximate solutions of linear shell problems. In a number of papers (see e.g. [1]—[4]) the classification of the linear shell theories is given on the basis of the concept of the characteristic wave length of the stress and deformation pattern. This is accomplished by establishing the limits of validity of the different classical linear shell theories in terms of the characteristic wave length L and characteristics of geometry of the middle and edge surfaces of the shell. The role of the external actions included in the boundary conditions is recognized in these papers only trough the definition of L as being the inverse rate of the maximum variation of any of the given shell data. In this paper the application of the method of asymptotic integration in analysis of the equations of the geometrically nonlinear theory of thin shells yields results, which indicate the way in which information on quantitative properties of the external actions has to be taken into account when establishing the limits of validity of the classical linear and geometrically nonlinear shell theories. The procedure is proposed for estimating the influences of the different sorts of external actions on the stress and strain state of the shell and the wave length of the most influential external action is suggested to be chosen as the characteristics wave length L of the problem.

Our intention is not to dwell upon the detailed presentation of the admittedly very complicated records of systems of equations derived by the asymptotic procedure. Rather, we focus our efforts on collecting the basic information resulting from the asymptotic analysis, which can give us a satisfactory picture of the shell behaviour even before the numerical solution of the problem starts.

The plan of the paper is as follows: Section 2 contains the collection of the basic equations and boundary conditions of the threedimensional problem and presents the main features of the method of asymptotic integration. The formal description of the way in which the asymptotic approximation to the exact solution may be constructed will be shown in Section 3 with special attention given to the presentation of the matching procedure whereby the different functions are combined to satisfy the edge conditions. Once the essence of the asymptotic analysis is given, the details can be found in [5].

In Section 4 the classification of the basic shell problems is given with the conditions limiting the ranges of validity of the corresponding classical twodimensional theories and the procedure for estimating influencies

[1]) The abbreviated version of this paper was delivered to the 13th IUTAM Congress 1972 in Moscow.

of the external actions is indicated. The predictions of the shell behavior made on the basis of the asymptotic analysis are compared with the results of the classical numerical analysis for some examples in Section 5 concluded with comments on the results.

2. Formulation. The method of asymptotic integration in general

In what follows Ω will denote the domain of the threedimensional space occupied by the shell in its undeformed state. The points of this domain will be identified by a set of material curvilinear coordinates ϑ^i $(i = 1, 2, 3)$ which in the undeformed state constitute a system of normal coordinates, the distance from the middle surface $\vartheta^3 = 0$ being measured by ϑ^3 coordinate. Coordinates ϑ^α represents the GAUSSian coordinates on the middle surface with corresponding covariant derivatives being denoted by symbols D_α. Symbol D_3 represents then a partial derivative with respect to the undeformed ϑ^3 coordinate. The boundary $\partial\Omega$ of the shell is formed by the faces $\partial\Omega^S = \partial\Omega^S_{(+)} \cup \partial\Omega^S_{(-)}$ with the upper face $\partial\Omega^S_{(+)}$ and the lower face $\partial\Omega^S_{(-)}$, and by the edge surface $\partial\Omega^B$, which will be assumed to be the coordinate surface $\vartheta^1 = 0$.

In our further consideration, the metric tensor $a_{\alpha\beta}$ on the middle surface will be assumed to satisfy $a_{\alpha\beta} \approx 1$, the curvature tensor of this surface will be denoted by $b_{\alpha\beta}$ and f_{ijkl} will be the tensor of elastic coefficients of the material of which the shell is made.

The system of the fundamental equations of the geometrically nonlinear theory of thin shells consists of three equations of the statical equilibrium and six equations of the generalized HOOKE's law. The nine basic unknowns of this formulation are the six components s^{ij} of the symmetric second PIOLA-KIRCHHOFF's stress tensor and three components u_i of the displacement vector $\boldsymbol{V} = u_i \boldsymbol{a}^i$, where $\boldsymbol{a}^i$ are contravariant base vectors associated with undeformed coordinates ϑ^i on the middle surface of the shell.

For our purpose, the basic nonlinear equations suffice to be referred to only in the general condensed form

$$\vartheta^i \in \Omega : g^E(u_i, D_j u_i, D_j D_k u_i, s^{ij}, D_k s^{ij}, b_{\alpha\beta}, D_\gamma b_{\alpha\beta}, f_{ijkl}, \vartheta^3) = 0 \,, \tag{2.1}$$

where g^E is a polynomial in its variables [5]. Unless otherwise stated, the Roman indices take the values 1, 2, 3 and the Greek indices the values 1, 2.

On the faces $\partial\Omega^S$ the external load is prescribed with components q^i

$$\vartheta^i \in \partial\Omega^S : g^S(s^{ij}, q^i) = 0 \,, \tag{2.2}$$

and on the edge $\partial\Omega^B$ (here given by $\vartheta^1 = 0$), the boundary conditions may be given in terms of stresses, displacements or in a mixed form

$$\vartheta^i \in \partial\Omega^B : g^B(u_i, D_j u_i, s^{ij}, t^i, v_i) = 0 \,. \tag{2.3}$$

The components of the external edge stress actions are denoted by t^i, the prescribed displacements by v_i.

In (2.2), (2.3), the functions g^S and g^B are also polynomials in the variables indicated. It is assumed that they are linear with respect to the external actions terms q^i, t^i and v_i. In addition, all the data functions of the considered shell problem are supposed to be sufficiently smooth. When dealing with expressions (2.1) to (2.3) simultaneously, the more condensed notation is used conveniently

$$\begin{aligned} &g(x, Dx, y, z) = 0 \,, \\ &x = \{u_i, s^{ij}\} \,, \qquad Dx = \{D_j u_i, D_j D_k u_i, D_k s^{ij}\} \,, \\ &y = \{q^i, t^i, v_i\} \,, \qquad z = \{b_{\alpha\beta}, D_\gamma b_{\alpha\beta}, f_{ijkl}, \vartheta^3\} \end{aligned} \tag{2.4}$$

denoting the set of the basic unknowns by symbol x, the set of their covariant derivatives by Dx and the sets of the coefficients and the external actions by z and y, respectively.

In the method of asymptotic integration, the approximate solution Ax of the problem (2.4) is assumed to have a form of a sum of several different approximating functions ${}_{(J)}x$

$$x = Ax = \sum_J {}_{(J)}x \,, \qquad J \in \mathfrak{J} \,. \tag{2.5}$$

Each of the approximating functions ${}_{(J)}x$ is constructed as a main part of the approximate solution $A\,x$ within its own region of approximation $\Omega^{(J)}$. The order in which the functions ${}_{(J)}x$ are subsequently determined induces the corresponding ordering in the summation set $\mathfrak{J}$.

To the minimal element of the set $\mathfrak{J}$, an approximating function will correspond which is to be determined in the first stage of the forming of the approximate solution (2.5). This function will be called the fundamental approximating function, it will approximate the solution of (2.4) in some sense in the whole domain Ω and will be assumed to possess an isotropic variation pattern over middle surface coordinates ϑ^α

$$\Omega^{(J)} = \Omega \,, \qquad L^{(J)}_\alpha = L \,. \tag{2.6}$$

By $L^{(J)}_i$ we denote the characteristic wave lengths of the stress and displacement pattern

$$L^{(J)}_i = \frac{\sup |{}_{(J)}x|}{\sup |D_i\, {}_{(J)}x|} \qquad \text{or} \qquad [D_i\, {}_{(J)}x] \approx \frac{{}_{(J)}x}{L^{(J)}_i} \,. \tag{2.7}$$

In what follows only thin shells will be considered with $L_3^{(J)} \approx h$, h being a measure of the thickness of the shell. From (2.7) starting, the summation will be always indicated by the usual summation symbol, therefore, there is no summation in (2.7).

In the edge zone (and also in the vicinity of the discontinuities of the shell data, if such is the case), the excess (or supplementary) approximating functions are constructed and combined with the fundamental approximating function in order to approximate the exact solution of (2.4). For their regions of approximation $\Omega^{(J)}$ we have

$$\Omega^{(J)} = \{\vartheta^i \in \Omega : \text{dist}\,(\vartheta^i, \partial\Omega^B) \leqq L_1^{(J)}\}, \tag{2.8}$$

which represent the boundary strips of the width equal to the corresponding wave length $L_1^{(J)}$.

The variation patterns of these functions are non-isotropic

$$L_1^{(J)} \ll L_2^{(J)} \qquad (L_2^{(J)} \approx L)\,, \tag{2.9}$$

and the extreme values of these functions are required to be reached within the corresponding regions of approximation

$$\sup_{\vartheta^i \in \Omega^{(J)}} |_{(J)}x| \gg \sup_{\vartheta^i \notin \Omega^{(J)}} |_{(J)}x|\,. \tag{2.10}$$

Decreasing rapidly inwards the interior of the shell, the excess functions will not degrade the order of approximation of the solution (2.5) outside of their regions of approximation $\Omega^{(J)}$. The introduction of the concept of the characteristic wave lengths is a distinctive feature of the method of asymptotic integration. It allows the covariant and partial derivatives of the basic unknown functions to be estimated when performing the order of magnitude analysis of the problem (2.4). It is this point which is often considered to imply some starting assumption on the class of solution. In this paper, henceforth, the wave length L from (2.6) will be considered as one of the two basic parameters in developing the asymptotic analysis (the second one $\mathcal{P}$ will be introduced later on). Depending on the possible relations between two basic parameters and the characteristics of the shell data, the basic types of the approximate solution (2.5) will be presented. Then the values of the basic parameters L, $\mathcal{P}$ will be specified in such a way, as to meet the properties of the boundary data by the particular type of the approximate solution (2.5).

3. Asymptotic integration procedure

In this Section, the main steps in the determination of the approximating functions $_{(J)}x$ from (2.5) will be given, references for the details being made to [5], [6].

When an approximating function with index $J = J^*$ is to be determined, all the functions with indices $J < J^*$ are supposed to be set up in the sense that the procedures were established allowing these functions to be found with the required asymptotic accuracy.

The first step in developing the method of asymptotic integration is a proper transformation of the basic equations (2.4), which consists of the adequate scaling of the coordinates ϑ^i. When the excess approximating functions are set up, then this transformation comprises also the transcription of the basic equations (2.4) into the local coordinate system near the edge, with ϑ^1 coordinate being the coordinate in the normal direction to the edge surface. In performing this transcription, the coefficients z from (2.5) are expanded in TAYLOR series near the edge surface $\vartheta^1 = 0$ and the covariant derivatives are split into terms containing partial derivatives and terms comprising products of unknown functions and the TAYLOR series of the CHRISTOFFEL symbols $\Gamma^\alpha_{\beta\gamma}$. Denoting this composed transformation by $_{(J)}\mathsf{T}$ we can characterize it by writing

$$_{(J)}\mathsf{T}\colon \theta^\alpha_{(J)} = \frac{\vartheta^\alpha}{L_\alpha^{(J)}}, \qquad \theta^3 = \theta^3_{(J)} = \frac{\vartheta^3}{h}, \qquad D_\alpha^{(J)} = L_\alpha^{(J)}\, D_\alpha\,, \qquad D_3^{(J)} = h\, D_3 = h\, \partial_3\,, \tag{3.1}$$

and in case $J > J_{\min}$ also by

$$D_\alpha^{(J)} = L_\alpha^{(J)}\, \partial_\alpha\,,$$

$$z_\mathfrak{z} = \sum_{r=0} \frac{1}{r!} (\vartheta^1)^r\, (\partial_1^r z_\mathfrak{z})_{\vartheta^1=0}\,,$$

$$z_\mathfrak{z} \in \{b_{\alpha\beta}, f_{ijkl}, \Gamma^\alpha_{\beta\gamma}\}\,,$$

where $\theta^i_{(J)}$ are new reduced coordinates, $D_i^{(J)}$ represent the derivatives in this new coordinate system, ∂_i are partial derivatives in coordinates ϑ^i, $z_\mathfrak{z}$ is any individual element of the set indicated above and h is the characteristic thickness of the shell.

Substituting

$$A^{(J)}x = {}_{(J)}x + \sum_{J^*<J} \sum_{r=0} \frac{1}{r!} (\vartheta^1)^r\, (\partial_1^r\, {}_{(J^*)}x)_{\vartheta^1=0}\,, \tag{3.2}$$

for the exact solution x into the transformed equations (2.4), we obtain the governing equations for determination of the approximating function $_{(J)}x$

$$_{(J)}g({}_{(J)}x, {}_{(J)}Dx, {}_{(J)}y, {}_{(J)}z) = \{{}_{(J)}\mathsf{T}[g(x, Dx, y, z)]\}_{x=A^{(J)}x} = 0\,. \tag{3.3}$$

If the excess approximating function is to be found ($J > J_{\min}$) then the list of coefficients ${}_{(J)}z$ in (3.3) differs from that of (2.4)

$$
\begin{aligned}
{}_{(J)}z &= \{[\partial_1^r z_\delta^*]_{\vartheta^1=0}, \vartheta^3, \vartheta^1\}, \qquad r = 0, 1, \ldots, \\
z_\delta^* &\in \{b_{\alpha\beta}, f_{ijkl}, \Gamma^\alpha_{\beta\gamma}, {}_{(J^*)}x\}, \qquad J^* < J,
\end{aligned}
\tag{3.4}
$$

since in (3.2) the edge values of the approximating functions constructed previously are considered to be known, too.

When constructing the governing equations for the determination of the approximating functions, the edge conditions (2.3) play rather a role of an indicator showing whether the number and properties of the approximating functions already at the hand are sufficient for satisfying all conditions on the edge surface. These conditions come fully in the action in the matching procedure, which will be described at the end of the present Section, and which results in specifying boundary conditions for (3.3). The excess approximating functions are required by (2.10) to fulfil also some decaying conditions, which will be assumed here to be

$$\{[(\theta^1_{(J)})^m \, (D_1^{(J)})^n \, {}_{(J)}x]\} \to 0 , \qquad \text{for} \qquad \theta^1_{(J)} \to \infty , \qquad n, m = 0, 1, \ldots \tag{3.5}$$

In analyzing equations (3.3) it is, therefore, necessary to establish also the existence conditions of the exponentially decaying solution.

Now the equations (3.3) are subjected to the asymptotic analysis. Each term φ entering the list of variables of function ${}_{(J)}g$ from (3.3) is expressed in the form

$$\varphi = \mathcal{M}(\varphi)\, N(\varphi) , \qquad \varphi \in \{{}_{(J)}x, {}_{(J)}y, {}_{(J)}z\} , \qquad \mathcal{M}(\varphi) = \sup_{\Omega_\varphi} |\varphi| , \qquad N(\varphi) = O(1) . \tag{3.6}$$

The quantity $\mathcal{M}(\varphi)$ represents the order of magnitude scalar characteristics of the term φ and $N(\varphi)$ is the nondimensional counterpart of this term (Ω_φ is the region of definition of the function φ). Using now the diameter $\mathcal{D}$ of the region Ω as the reference length and introducing the second basic parameter $\mathcal{P}$ as the reference stress, all the order of magnitude characteristics $\mathcal{M}(\varphi)$ of terms φ, which enter the equations (3.3), are reduced to the corresponding reference measures. Declaring these ratios as powers of the parameter η, the nondimensional form of (3.3) takes, finally, the form of the system of equations with small parameter η^{-1}

$$
\begin{aligned}
&{}_{(J)}g({}_{(J)}x, {}_{(J)}Dx, {}_{(J)}y, {}_{(J)}z) = {}_{(J)}g^N({}_{(J)}X, \ldots, \eta^{-1}) = 0 , \\
&{}_{(J)}X = N({}_{(J)}x) ,
\end{aligned}
\tag{3.7}
$$

denoting conveniently the nondimensional counterpart of the term x by the corresponding capital letter X. The parameter η^{-1} is introduced by

$$\eta^{-q} = \frac{h}{\mathcal{D}} \qquad (q > 0) , \tag{3.8}$$

and for thin shells it is, actually, a small number. From the list of the ratios entering (3.7), the following ones are important

$$\frac{R}{\mathcal{D}} = \eta^{p^+} , \qquad \frac{L}{\mathcal{D}} = \eta^{-p} , \qquad \frac{E}{\mathcal{P}} = \eta^{-\varrho} , \qquad \frac{L_\alpha^{(J)}}{\mathcal{D}} = \eta^{-p_\alpha^{(J)}} , \qquad R = \frac{1}{\mathcal{M}(b_{\alpha\beta})}, \qquad E = \frac{1}{\mathcal{M}(f_{ijkl})}, \tag{3.9}$$

where R is the characteristic curvature radius and E is the characteristic elastic modulus of the shell problem.

The exponents q, p^+, relating basic geometrical characteristics of the shell to the reference length $\mathcal{D}$, will be called basic exponents. Parametric exponents are then p and ϱ, which relate basic parameters L and $\mathcal{P}$ to the shell data characteristics D and E respectively.

In governing equations (3.7) the following ratios appear also

$$\frac{\mathcal{M}({}_{(J)}u_i)}{\mathcal{D}} = \eta^{\varkappa^{(J)} + r_i^{(J)}} , \qquad \frac{\mathcal{M}({}_{(J)}s^{ij})}{\mathcal{P}} = \eta^{\varkappa^{(J)} + r_{ij}^{(J)}} , \tag{3.10}$$

with $r^{(J)}$ being called the unknown exponents and $\varkappa^{(J)}$ playing the role of the matching exponents, which are to be determined when matching different approximating functions ${}_{(J)}x$ in order to satisfy the edge conditions. In the transformed form of the surface and of the edge conditions we find also the ratios

$$\frac{\mathcal{M}(q^i)}{\mathcal{P}} = \eta^{q_i^*}, \qquad \frac{\mathcal{M}(t^i)}{\mathcal{P}} = \eta^{t_i^*}, \qquad \frac{\mathcal{M}(v_i)}{\mathcal{D}} = \eta^{v_i^*}, \tag{3.11}$$

with exponents q_i^*, t_i^* and v_i^* which will be named the external actions exponents.

Next, the order of magnitude analysis of (3.7) is performed to find interrelations between the basic, the parametric, the unknown and the external actions exponents from (3.9)—(3.11), which allow the system of equations (3.7) with small parameter η^{-1} to be solved by using asymptotic expansions of the nondimensional counterparts of the unknowns

$${}_{(J)}X = \sum_{s \geqq 0} \eta^{-s} \, {}_{(J)}X^{[s]} . \tag{3.12}$$

These interrelation have generally the form

$$\begin{aligned} &1.\ C(q, p^+, p, \varrho) \leqq 0 ;\\ &2.\ r^{(J)} = r^{(J)}(q, p^+, p, \varrho) , \quad p_1^{(J)} = p_1^{(J)}(q, p^+, p, \varrho) ;\\ &3.\ \varkappa^{(J)} \leqq k^{(J)}(q, p^+, p, \varrho) ; \end{aligned} \tag{3.13}$$

and can be found in [5]. Here, we present only one example case corresponding to the fundamental approximating function whose lowest order coefficients from (3.12) can be determined by solving equations of the bending shell theory:

$$\begin{aligned} &1.\ p^+ \geqq q - 2\,p , \quad \varrho \geqq 4\,(q - p) ;\\ &2.\ r_{33}^{(J)} = 0 , \quad r_{\alpha 3}^{(J)} = q - p , \quad r_{\alpha\beta}^{(J)} = 2\,(q - p) , \quad r_{\alpha}^{(J)} = 2\,q - 3\,p - \varrho ,\\ &\quad r_3^{(J)} = 3\,q - 4\,p - \varrho ;\\ &3.\ \varkappa^{(J)} = 0 . \end{aligned} \tag{3.14}$$

After inserting (3.12) and (3.13.2) in (3.7) and after equating the same powers of the parameter η^{-1} in (3.7) we obtain iteration processes for successive determination of the coefficients of expansions (3.12) in form

$$\begin{aligned} &{}_{(J)}g^{[s]}({}_{(J)}X^{[s]}, {}_{(J)}DX^{[s]}, {}_{(J)}Z^{[s]}, \ldots) = 0 ,\\ &{}_{(J)}DX^{[s]} = \{D_i^{(J)}\, {}_{(J)}X^{[s]}, D_i^{(J)}\, D_j^{(J)}\, {}_{(J)}X^{[s]}\} ,\\ &{}_{(J)}Z^{[s]} = \{{}_{(J)}X^{[s^*]}, \ldots\} , \quad s^* < s . \end{aligned} \tag{3.15}$$

Here once more only the decisive terms of the list of variables of the polynomial function ${}_{(J)}g^{[s]}$ are presented, those which can give us basic idea on the structure of the iteration process (3.15). Also in the list of elements of the nondimensional coefficients set ${}_{(J)}Z^{[s]}$ only those terms are given indicating the successive character of the process of determination of the coefficients ${}_{(J)}X^{[s]}$.

With the exception of the excess iteration process denoted by the summation index $J = D$, all the other iteration processes presented here have (3.15) in form allowing an easy integration with respect to θ^3 variable. Then, for the coefficients of expansions (3.12) we can write

$$ {}_{(J)}X_{\mathfrak{z}}^{[s]} = \sum_{n=0}^{\lambda} (\theta^3)^n\, {}_{(J)}X_{\mathfrak{z}}^{[s,n]} , \tag{3.16}$$

$$\lambda = \lambda(J, s, \mathfrak{z}) , \quad \mathfrak{z} \in \mathfrak{Z} , \quad \mathfrak{Z} = \{i, i\,j\} ,$$

where we introduce a set $\mathfrak{Z}$ of indices, with the aid of which we can point out any particular element of the set X. Coefficients of the polynomials (3.16) are successively determined from equations, which are less dimensional than (3.15)

$$\begin{aligned} &{}_{(J)}g^{[s,n]}({}_{(J)}X^{[s,n]}, D_\alpha^{(J)}\, {}_{(J)}X^{[s,n]}, \ldots, {}_{(J)}Z^{[s,n]}, \ldots) = 0 ,\\ &{}_{(J)}Z^{[s,n]} = \{{}_{(J)}X^{[s^*,n^*]}, \ldots\} , \quad s^* \leqq s , \quad n^* \leqq n , \quad s^* + n^* < s + n . \end{aligned} \tag{3.17}$$

On the s-th step only a few of the equations (3.17) represent actually a system of partial differential equations, which will be called the main system of equations of the corresponding iteration system and will serve for the determination of the so called basic expansions coefficients ${}_{(J)}\overset{*}{X}{}^{[s,n]}$ introduced when rewriting (3.16) in form

$$ {}_{(J)}X_{\mathfrak{z}}^{[s]} = \sum_{n=N_1}^{N_2} (\theta^3)^n\, {}_{(J)}\overset{*}{X}{}_{\mathfrak{z}}^{[s,n]} + \sum_{n=0}^{\lambda} (\theta^3)^n\, {}_{(J)}\widetilde{X}_{\mathfrak{z}}^{[s,n]} , \tag{3.18}$$

$$N_\alpha = N_\alpha(J, \mathfrak{z}) , \quad \alpha = 1, 2 , \quad \mathfrak{z} \in \mathfrak{Z} .$$

Here, the coefficients ${}_{(J)}\widetilde{X}^{[s,n]}$ are those which can be determined from the rest of (3.17) before the main system of equations is solved on the given step of the iteration process.

The representation (3.18) proves to be very useful when performing the matching of different approximating functions ${}_{(J)}x$ from (2.5) in order to satisfy edge conditions (2.3). Before describing the matching procedure in a more detailed way it will be expedient to summarize here some basic properties of the iteration process distinguished by index $J = D$. The excess approximating function determined by this process has to be incorporated in any type of solution (2.5), since only this process enables us to satisfy the edge conditions with the general form of the external actions t^i, v_i with respect to variable ϑ^3, the other iteration processes yielding approximating functions of the polynomial type (3.18).

The main system of equations of this excess iteration process is twodimensional in local coordinates $\theta_{(J)}^1$, θ^3. The conditions of the existence of the exponentially decaying solution of this problem are found in [7], [8] to have the form

$$\theta_{(J)}^1 = 0 : \int_{-1/2}^{1/2} \mathcal{F}_k({}_{(D)}X^{[s]}, \theta^3, f_{ijkl}, h)\, d\theta^3 = 0 , \qquad k = 1, 2, 3, 4 , \tag{3.19}$$

where $\mathcal{F}_k$ are polynomials in variables indicated (they are linear with respect to ${}_{(D)}X^{[s]}$). In the matching procedure one of the principles governing the finding of proper values of the matching exponents $\varkappa^{(J)}$ is the requirement that conditions (3.19) are to be satisfied by the combination of the iteration processes with $J < D$.

Passing on to the description of the matching procedure we confine our attention to the edge condition of the form

$$\vartheta^1 = 0 : x_{\mathfrak{z}} = y_{\mathfrak{z}} + \cdots , \qquad \mathfrak{z} \in \mathfrak{Z}_B , \qquad \mathfrak{Z}_B \subset \{i, i\,1\} , \tag{3.20}$$

where by points we indicate terms which in the asymptotic analysis show to be of a smaller order of magnitude than $x_{\mathfrak{z}}$. The set $\mathfrak{Z}_B$ represents three-element-subset of the index set $\{i, i\,1\}$ and its structure depends on the type of the edge conditions, e.g. for the free edge we have $\mathfrak{Z}_B = \{11, 21, 31\}$, for the hinged edge the version with $\mathfrak{Z}_B = \{11, 2, 3\}$ can be considered and the clamped edge is characterised by $\mathfrak{Z}_B = \{1, 2, 3\}$.

It seems to be quite superfluous to make (3.20) more clear by

$$\begin{aligned} &\mathfrak{z} = i \quad : x_{\mathfrak{z}} = u_i , && y_{\mathfrak{z}} = v_i , && \chi_{\mathfrak{z}} = v_i^* , \\ &\mathfrak{z} = i\,1 : x_{\mathfrak{z}} = s^{i1} , && y_{\mathfrak{z}} = t^i , && \chi_{\mathfrak{z}} = t_i^* , \\ &(\mathfrak{z} = i\,3 : x_{\mathfrak{z}} = s^{i3} , && y_{\mathfrak{z}} = q^i , && \chi_{\mathfrak{z}} = q_i^*) \end{aligned} \tag{3.21}$$

with $\chi_{\mathfrak{z}}$ being further used to represent the external actions exponents from (3.11).

Now we split the external action terms $y_{\mathfrak{z}}$ in two parts of which the first ones represent the polynomials in θ^3 variable. Then if possible and expedient we expand both the parts of these decompositions in asymptotic series with respect to powers of the small parameter η^{-1} and using (3.6), (3.10) to (3.12), (3.16), (3.18) and (3.21) we are, finally, in a position to write the edge conditions (3.20) in form convenient for accomplishing the matching procedure

$$\begin{aligned} &\sum_{J<D} \left[\eta^{\varkappa^{(J)} + r_{\mathfrak{z}}^{(J)}} \left(\sum_{s \geq 0} \eta^{-s} \sum_{n=N_1}^{N_2} (\theta^3)^n \, {}_{(J)}\overset{*}{X}{}_{\mathfrak{z}}^{[s,n]} \right) \right] + \eta^{\varkappa^{(D)} + r_{\mathfrak{z}}^{(D)}} \sum_{s \geq 0} \eta^{-s} \, {}_{(D)}X_{\mathfrak{z}}^{[s]} \\ &= \eta^{\chi_{\mathfrak{z}}} \left[\left(\sum_{n=0}^{N^*} (\theta^3)^n \sum_{s \geq \chi^n} \eta^{-s} \, Y_{\mathfrak{z}}^{[s,n]} \right) + \sum_{s \geq \chi^*} \eta^{-s} \, \overset{*}{Y}{}_{\mathfrak{z}}^{[s]} \right] - \sum_{J<D} \left[\eta^{\varkappa^{(J)} + r_{\mathfrak{z}}^{(J)}} \left(\sum_{n=0} (\theta^3)^n \sum_{s \geq \psi} \eta^{-s} \, {}_{(J)}\widetilde{X}_{\mathfrak{z}}^{[s,n]} \right) \right] + \cdots , \\ &N^* = \max_J \{N_2(J, \mathfrak{z})\} , \qquad \mathfrak{z} \in \mathfrak{Z}_B . \end{aligned} \tag{3.22}$$

For the sake of better legibility in (3.22) no indications are made of the dependence of the exponents χ^n, χ^* and ψ on the variables J, $\mathfrak{z}$ or n.

These exponents are defined by

$$\begin{aligned} \chi^n &= \chi^n(\mathfrak{z}, n) = \min \{s : Y_{\mathfrak{z}}^{[s,n]} \neq 0\} , \\ \chi^* &= \chi^*(\mathfrak{z}) = \min \{s : \overset{*}{Y}{}_{\mathfrak{z}}^{[s]} \neq 0\} , \\ \psi &= \psi(J, \mathfrak{z}, n) = \min \{s : {}_{(J)}\widetilde{X}_{\mathfrak{z}}^{[s,n]} \neq 0\} , \end{aligned} \tag{3.23}$$

indicating the order numbers of the first nonvanishing (tilded) coefficients of expansions (3.18), and of the external actions expansions

$$\frac{1}{\mathcal{X}} y_{\mathfrak{z}} = \eta^{\chi_{\mathfrak{z}}} \left(\sum_{n=0}^{N^*} (\theta^3)^n \sum_{s \geq 0} \eta^{-s} \, Y_{\mathfrak{z}}^{[s,n]} + \sum_{s \geq 0} \eta^{-s} \, \overset{*}{Y}{}_{\mathfrak{z}}^{[s]} \right), \tag{3.24}$$

$$\mathfrak{z} \in \mathfrak{Z} , \qquad \mathcal{X} = \mathcal{D} \quad \text{if} \quad \mathfrak{z} = i , \qquad \mathfrak{X} = \mathcal{P} \quad \text{if} \quad \mathfrak{z} = i\,1 .$$

From (3.22) it follows that the line of inequalities should be satisfied by a proper specification of the values of the matching exponents $\varkappa^{(J)}$

$$\begin{aligned} &1. \text{ for } J < D : \varkappa^{(J)} + r_{\mathfrak{z}}^{(J)} \geq \chi_{\mathfrak{z}} - \chi^n(\mathfrak{z}, n) , \\ &\qquad \varkappa^{(J)} + r_{\mathfrak{z}}^{(J)} \geq \varkappa^{(J^*)} + r_{\mathfrak{z}}^{(J^*)} - \psi(J^*, \mathfrak{z}, n) , \\ &\text{when } \mathfrak{z} \in \mathfrak{Z}_B , \qquad N_1(J, \mathfrak{z}) \leq n \leq N_2(J, \mathfrak{z}) , \qquad J^* < D \quad (J^* \neq J) , \\ &2. \text{ for } J \leq D : \varkappa^{(J)} + r_{\mathfrak{z}}^{(J)} \geq \chi_{\mathfrak{z}} - \chi^*(\mathfrak{z}) , \\ &3. \text{ for } J^* < D : \varkappa^{(D)} + r_{\mathfrak{z}}^{(D)} \geq \varkappa^{(J^*)} + r_{\mathfrak{z}}^{(J^*)} - \psi(J^*, \mathfrak{z}, N^* + 1) . \end{aligned} \tag{3.25}$$

To fulfil (3.25) the matching exponents $\varkappa^{(J)}$ will be determined in a succession according to

$$\begin{aligned} &\varkappa^{(J)} = \max_{\mathfrak{z}, J^* < J} \{\chi_{\mathfrak{z}} - \tilde{\chi}_{\mathfrak{z}} - r_{\mathfrak{z}}^{(J)}, \varkappa^{(J^*)} + r_{\mathfrak{z}}^{(J^*)} - r_{\mathfrak{z}}^{(J)} - \tilde{\psi}_{\mathfrak{z}}\} , \\ &\tilde{\chi}_{\mathfrak{z}} = \min_n \{\chi^n(\mathfrak{z}, n), \chi^*(\mathfrak{z})\} , \\ &\tilde{\psi}_{\mathfrak{z}} = \min \{\psi(J^*, \mathfrak{z}, n)\} , \\ &\mathfrak{z} \in \mathfrak{Z}_B , \qquad N_1(J, \mathfrak{z}) \leq n \leq N_2(J, \mathfrak{z}) . \end{aligned} \tag{3.26}$$

At the same time, inequalities (3.25) require the satisfaction of

$$\begin{aligned} &\varkappa^{(J)} \leq \varkappa^{(J^*)} + r_{\mathfrak{z}}^{(J^*)} + \bar{\psi}_{\mathfrak{z}} - r_{\mathfrak{z}}^{(J)} \quad \text{for} \quad J^* < J , \quad \mathfrak{z} \in \mathfrak{Z}_B , \\ &\bar{\psi}_{\mathfrak{z}} = \max_n \{\psi(J, \mathfrak{z}, n)\} \qquad \text{for} \quad N_1(J^*, \mathfrak{z}) \leq n \leq N_2(J^*, \mathfrak{z}) . \end{aligned} \tag{3.27}$$

When exponents $\varkappa^{(J)}$ are established on the basis of (3.26) and (3.27), the boundary conditions for the iteration processes follow from (3.22) by equating the coefficients of the same powers of η^{-1}. The equations derived, which contain no ${}_{(D)}X^{[s]}$ terms, serve immediately as boundary conditions for the main system of equations for one of the iteration processes with $J < D$. When also terms ${}_{(D)}X^{[s]}$ are present, then the existence conditions (3.19) come in operation and yield four boundary conditions for the iteration processes $J < D$.

From (3.22) we have

$$ {}_{(D)}X_{\mathfrak{z}}^{[s]} = {}_{(D)}B_{\mathfrak{z}}^{[s]}\big({}_{(J)}\overset{*}{X}{}_{\mathfrak{z}}^{[s^*, n^*]}, \ldots\big) , \qquad \mathfrak{z} \in \mathfrak{Z}_B , \quad J < D , \quad s^* \geqq s , \tag{3.28} $$

and then the inserting of (3.28) into (3.19) results in four boundary conditions for the iteration processes with $J < D$

$$ {}_{(J)}B_{i}^{[s]}\big({}_{(J)}\overset{*}{X}{}_{\mathfrak{z}}^{[s^*, n^*]}, \ldots\big) = 0 , \qquad J < D , \qquad i = 1, 2, 3, 4 . \tag{3.29} $$

The satisfaction of these latter conditions allows the boundary conditions (3.28) of the iteration process $J = D$ to meet all the required existence conditions (3.19).

For the fundamental iteration process (to which corresponds minimal element of the set $\mathfrak{J}$ in (2.5)) it is convenient to put

$$ \varkappa^{(J)} = 0 , \qquad J = J_{\min} , \tag{3.30} $$

and then from (3.25) it follows

$$ r_{\mathfrak{z}}^{(J)} \geqq \chi_{\mathfrak{z}}^{(J)} , \qquad \mathfrak{z} \in \mathfrak{Z}_E , \quad \mathfrak{Z}_E = \mathfrak{Z}_B \cup \{i\,3\} , \tag{3.31} $$

when also results of the asymptotic analysis of the surface conditions were taken into account. Having a look at (3.21), (3.9) to (3.11) and (3.13), the inequalities (3.31) show us a possibility of expressing the second basic parameter $\mathscr{P}$ (introduced as a reference stress) in terms of the shell data characteristics and the characteristic wave length L, but we find it more convenient to present these results in a separate Section.

4. Classification of the classical linear and geometrically nonlinear shell theories

In this Section we present and discuss a few basic types of the approximate solution (2.5) of the presented shell problem (2.4). The shell problems with special properties of the middle and edge surfaces will not be included in our considerations. The basic types of the solution (2.5) discussed here correspond to the following interrelations between the basic and the parametric exponents introduced in (3.9)—(3.11) (these interrelations are expressed also in terms of the shell characteristics h, R, E and of the basic parameters L, $\mathscr{P}$)

$$ \begin{array}{lll} 1. & p^+ < q - 2p & : \; hR \ll L^2 , \\ & \varrho \geqq 2(q + p^+) & : \; \mathscr{P} \lessapprox E\,h^2/R^2 \;;^{1)} \\ 2. & p^+ < q - 2p & : \; hR \ll L^2 , \\ & q + 2p + 3p^+ < \varrho < 2(q + p^+) & : \; E\,h^2/R^2 \ll \mathscr{P} \ll E\,h\,L^2/R^3 \;; \\ 3. & p^+ \geqq q - 2p & : \; hR \gtrapprox L^2 , \\ & \varrho > 4(q - p) & : \; \mathscr{P} \ll E\,h^4/L^4 \;; \\ 4. & p^+ \geqq q - 2p & : \; hR \gtrapprox L^2 , \\ & \varrho = 4(q - p) & : \; \mathscr{P} \approx E\,h^4/L^4 \;; \\ 5a. & p^+ < q - 2p & : \; hR = L^2 , \\ & q - p < \varrho < q + 2p + 3p^+ & : \; E\,h\,L^2/R^3 \ll \mathscr{P} \ll E\,h/L \;; \\ 5b. & p^+ \geqq q - 2p & : \; hR \gtrapprox L^2 , \\ & q - p < \varrho < 4(q - p) & : \; E\,h^4/L^4 \ll \mathscr{P} \ll E\,h/L . \end{array} \tag{4.1} $$

The corresponding sets of non-contradictory exponents "r" from (3.13) can be found in [5] and then (3.31) yields the definition of the second basic parameter $\mathscr{P}$ as the characteristics of the external actions.

In fact, denoting the order of magnitude characteristics of the external actions by

$$ Q_i = \mathscr{M}(q^i) , \qquad T_i = \mathscr{M}(t^i) , \qquad V_i = \mathscr{M}(v_i) , \tag{4.2} $$

we deduce from (3.30)

$$ \mathscr{P} = \max\,[Q_3, {}_{(J)}\mathscr{P}_{\alpha 3}, {}_{(J)}\mathscr{P}_{i1}, {}_{(J)}\mathscr{P}_i] . \tag{4.3} $$

Here the term Q_3 represents the influence of the normal surface load q^3 and ${}_{(J)}\mathscr{P}_{\alpha 3}$ influence of the tangential surface load q^α. The influences of the edge loads t^i are represented by ${}_{(J)}\mathscr{P}_{i1}$ and the terms ${}_{(J)}\mathscr{P}_i$ stand for the influences of the prescribed edge displacements v_i.

In cases (4.1.1) and (4.1.2) we put in (4.3) $J = A$, this index will be also ascribed to all quantities associated with corresponding fundamental iteration process (in (2.5) we have, therefore, $J_{\min} = A$ in set $\mathfrak{J}$). The

[1]) Composed symbol $\lessapprox$ means "$\approx$ or $<$"

main system of the partial differential equations of this iteration process will coincide in the form with the equations of the linear membrane shell theory and for $_{(J)}\mathcal{P}$ terms from (4.3) we have

$$\begin{aligned} {}_{(A)}\mathcal{P}_{\alpha 3} &= \frac{Q_\alpha L}{R}, \qquad {}_{(A)}\mathcal{P}_{\alpha 1} = \frac{T_\alpha h}{R}, \qquad {}_{(A)}\mathcal{P}_{31} = \frac{T_3 L}{R}, \\ {}_{(A)}\mathcal{P}_{\alpha} &= \frac{V_\alpha h E}{L R}, \qquad {}_{(A)}\mathcal{P}_{3} = \frac{V_3 h E}{R^2}. \end{aligned} \tag{4.4}$$

In the case (4.1.3), the main system of the fundamental iteration process corresponds to the system of equations of the classical linear bending theory of plates and shallow shells. For this case we write in (2.5) $J_{\min} = B\,L$, whereas the notation $J_{\min} = B\,N$ in (2.5) is used for case (4.1.4), the main system of which in the lowest order approximation represents the equations of the geometrically nonlinear plate and shallow shell theory. In both these cases we deduce from (3.30)

$$ {}_{(B)}\mathcal{P}_{\alpha 3} = \frac{Q_\alpha h}{L}, \quad {}_{(B)}\mathcal{P}_{\alpha 1} = \frac{T_\alpha h^2}{L^2}, \quad {}_{(B)}\mathcal{P}_{31} = \frac{T_3 h}{L}, \quad {}_{(B)}\mathcal{P}_{\alpha} = \frac{V_\alpha E h^2}{L^3}, \quad {}_{(B)}\mathcal{P}_{3} = \frac{V_3 E h^3}{L^4}. \tag{4.5}$$

Finally, we write $J_{\min} = C$ in case (4.1.5) and the lowest order main system of the corresponding fundamental iteration process coincides with equations of the classical nonlinear membrane theory. Here we have

$$\begin{aligned} {}_{(C)}\mathcal{P}_{\alpha 3} &= \frac{(Q_\alpha)^{3/2} L^{1/2}}{(h E)^{1/2}}, \qquad {}_{(C)}\mathcal{P}_{\alpha 1} = \frac{(T_\alpha)^{3/2} h}{E^{1/2} L}, \qquad {}_{(C)}\mathcal{P}_{31} = \frac{(T_3)^{3/2} L^{1/2}}{(h E)^{1/2}}, \\ {}_{(C)}\mathcal{P}_{\alpha} &= \frac{(V_\alpha)^{3/2} h E}{L^{5/2}}, \qquad {}_{(C)}P_{3} = \frac{(V_3)^3 h E}{L^4}. \end{aligned} \tag{4.6}$$

When speaking about the iteration processes in this Section we have in mind the procedures for the subsequent determinations of the coefficients of expansions

$$x_{\mathfrak{z}} \approx A\, x_{\mathfrak{z}} = \sum_{J<D} \sum_{s\geq 0} \sum_{n=0}^{\lambda} (\vartheta^3)^n \, {}_{(J)}x_{\mathfrak{z}}^{[s,n]} + \sum_{s\geq 0} {}_{(D)}x_{\mathfrak{z}}^{[s]}, \qquad \mathfrak{z} \in \mathfrak{Z}. \tag{4.7}$$

The correspondence between these coefficients and the coefficients used in preceding Sections can be established by comparing (4.7) with expression resulting from (2.5), (3.6), (3.12)

$$A\, x_{\mathfrak{z}} = \mathcal{X}\left[\sum_{J<D} \overset{*}{\eta}^{\varkappa(J)+r_{\mathfrak{z}}^{(J)}} \sum_{s\geq 0} \overset{*}{\eta}^{-s} \sum_{n\geq 0}^{\lambda} (\vartheta^3)^n \, {}_{(J)}X_{\mathfrak{z}}^{[s,n]} + \overset{*}{\eta}^{\varkappa(D)+r_{\mathfrak{z}}^{(D)}} \sum_{s\geq 0} \overset{*}{\eta}^{-s} \, {}_{(D)}X_{\mathfrak{z}}^{[s]}\right], \tag{4.8}$$

$$\mathfrak{z} \in \mathfrak{Z}, \qquad \mathcal{X} = \mathcal{D} \quad \text{if} \quad \mathfrak{z} = i, \qquad \mathcal{X} = \mathcal{P} \quad \text{if} \quad \mathfrak{z} = i\,j,$$

with $\overset{*}{\eta}^{-q}$ being now the particular value of the ratio $h/\mathcal{D}$ for the considered shell.

So far, we were concerned only with properties and notations of the fundamental iteration processes. To give an idea of the complete structure of the approximate solution (2.5) in five here considered cases (4.1), we write down the corresponding summation sets $\mathfrak{J}$ from (2.5), the ordering of the particular set $\mathfrak{J}$ being also indicated

$$\begin{aligned} &1.\ \mathfrak{J} = \{A < A\,B < D\}, \\ &2.\ \mathfrak{J} = \{A < A\,C\,n < A\,D\,n < D\}, \qquad n = 0, 1, \\ &3.\ \mathfrak{J} = \{B\,L < D\}, \\ &4.\ \mathfrak{J} = \{B\,N < D\}, \\ &5.\ \mathfrak{J} = \{C < C\,n < D\}, \qquad n = 0, 1, \ldots, \end{aligned} \tag{4.9}$$

and in following we shall present also the main equations of the excess iteration processes involved in (2.5).

The main features of the excess iteration process D were described in Section 3 and therefore, now we shall give attention only to the supplementary iteration processes denoted by indices $A\,B$, $A\,C\,n$, $A\,D\,n$ and $C\,n$. The familiar bending edge effect iteration process corresponds to index $A\,B$ with main equation

$$\partial_1^4 w + \alpha_1 w = W, \tag{4.10}$$

$$\alpha_1 = \left[\frac{12\,(b_{22})^2}{g^{1111}\, f_{2222}\, h^2}\right]_{\vartheta^1=0}, \qquad w = {}_{(AB)}u_3^{[s,0]},$$

where ∂_1 is symbol of the derivative with respect to ϑ^1 variable, W denotes the right side term and tensor $g^{\alpha\beta\gamma\varkappa}$ is the inverse of the tensor of elastic coefficients $f_{\alpha\beta\gamma\varkappa}$ $(g^{\alpha\beta\gamma\varkappa} f_{\gamma\varkappa\lambda\omega} = 1/2\,(\delta_\lambda^\alpha \delta_\omega^\beta + \delta_\omega^\alpha \delta_\lambda^\beta))$.

The existence condition $E\,1$ of this iteration process reads

$$E\,1\colon \quad \alpha_1 > 0 \to (b_{22})_{\vartheta^1=0} \approx \frac{1}{R}, \tag{4.11}$$

and requires the middle surface curvature not to vanish in the edge direction. When (4.11) is violated, then more special bending edge effects are to be considered. For the sake of the simplicity of our consideration these effects are not presented here.

In case (4.1.2), except the excess iteration process D, we have to use two more supplementary iteration processes ACn and ADn with main equations

$$ACn: \qquad (\vartheta^1)^n\, \partial_1^2 w - \frac{\alpha_1}{\alpha_2} w = W\,, \qquad w = {}_{(ACn)}u_3^{[s,0]}\,, \tag{4.12}$$

$$ADn: \qquad \partial_1^4 w - (\vartheta^1)^n \alpha_2\, \partial_1^2 w = W^*\,, \qquad n = 0, 1\,, \qquad w = {}_{(ADn)}u_3^{[s,0]}\,,$$

where we apply the notation already used in (4.10) and in addition

$$\alpha_2 = \left[\frac{12\,(\partial_1^n\, {}_{(A)}s_{[0,0]}^{11})}{n!\, h^2\, g^{1111}}\right]_{\vartheta^1=0}. \tag{4.13}$$

The existence conditions of these iteration processes are the condition $E\,1$ from (4.11) and condition $E\,2$ $(\alpha_2 > 0)$

$$[\partial_1^j\, {}_{(A)}s_{[0,0]}^{11}]_{\vartheta^1=0} = \delta_n^j s^*\,, \qquad s^* > 0\,, \qquad j = 0, 1, \ldots, n\,. \tag{4.14}$$

The condition $E\,2$ requires the *membrane* stress ${}_{(A)}s_{[0,0]}^{11}$ to satisfy some general positivness conditions on the edge surface. Here δ_n^j is $(n+1)$ dimensional KRONECKER's tensor.

In (4.14) when replacing index A by index C, we obtain the existence condition of the supplementary iteration process Cn $(n = 0, 1, 2, \ldots)$. The main equation of this process coincides with that of the iteration process derived in [9] (for $n = 0, 1$). We have

$$Cn: \qquad \partial_1^4 w - (\vartheta^1)^n \alpha_2\, \partial_1^2 w = W\,, \qquad w = {}_{(Cn)}u_3^{[s,0]}\,, \tag{4.15}$$

notation being that of (4.10)—(4.14) with replacements indicated above.

In a brief recapitulation it can be said that when basic characteristic quantities of the problem considered h, E, R, L, $\mathcal{P}$ satisfy relations (4.1.1), then in the interior of the shell the stress and deformation state prevailes which is proper to the linear membrane shell theory. The same situation occurs when external action characteristics $\mathcal{P}$ increases and takes values from (4.1.2), but in such case the edge value of the "membrane" stress should be positive

$$[{}_{(A)}s_{[0,0]}^{11}]_{\vartheta^1=0} > 0\,, \tag{4.16}$$

in order that approximate solution (2.5) may be constructed. The condition (4.16) (possibly in more general form (4.13)) comes in the force also in case (4.1.5), when the equations of the nonlinear membrane theory should be solved to obtain information on the lowest order approximation to the stress and strain state of the shell (4.1.5.a) or plate (4.1.5.b) considered.

Conditions (4.1.3) indicate cases, when the linear bending theory of plates and shallow shells can be used in forming the lowest order approximation; if external actions become larger in size so that $\mathcal{P}$ acquires values from (4.1.4), then the geometrically nonlinear bending theory of plates and shallow shells is to be used.

Now it is clear that value

$$\begin{aligned} \mathcal{P}_K &= E\,h^2/R^2 \quad \text{if} \quad h\,R \ll L^2\,, \\ \mathcal{P}_K &= E\,h^4/L^4 \quad \text{if} \quad h\,R \gtrapprox L^2\,, \end{aligned} \tag{4.17}$$

can be regarded as the critical value for external actions characteristics $\mathcal{P}$. If $\mathcal{P}$ exceeds this critical value, then the asymptotic procedure for the creating the solution (2.5) is conditioned by (4.13) (or in its simpler form by (4.16)). In case of non-shallow shells ($h\,R \ll L^2$), it is the linear membrane shell theory and in case of shallow shells and plates ($h\,R \geqq L^2$) the geometrically nonlinear bending theory which are to answer the question if conditions (4.13) are satisfied. These theories can be, therefore, regarded as theories which may actually handle the stability problems of shells and plates.

On the basis of the results contained in (4.3)—(4.6), the influence of the different kinds of the external actions can be appreciated and then the wave length of the prevailing external action is declared as the characteristic wave length of the problem considered. The details of this procedure, which results in establishing the values of the basic parameters L and $\mathcal{P}$, can be found in [13].

Moreover, rewriting (4.4)—(4.6) in the form

$${}_{(J)}\mathcal{P}_{\mathfrak{z}}\big(\mathcal{M}(x_{\mathfrak{z}}),\, L,\, h,\, R\big) = \mathcal{P}\,, \qquad \mathfrak{z} \in \mathfrak{Z} \tag{4.18}$$

the value of the basic parameter $\mathcal{P}$ being the *influence value* of the prevailing external action, the quantities $\mathcal{M}(x_{\mathfrak{z}})$ can be computed representing the order of magnitude information on the stresses and displacements. Such an information may be very valuable at the design stage, when often several possible cases are to be judged in order to find the optimal design solution.

5. Examples. Concluding remarks

In order to show that the procedure proposed here (see also [13]) for predicting the shell behaviour is rational, we inspect several cases of the isotropic plates and spherical shallow shells with square plan $d \times d$ as well as circular plates with radius r. The edges will be assumed hinged or clamped, the uniform loading will be of the normal surface type ($q^3 = \text{const}$) or of the normal edge type ($t^1 = \text{const}$). The selection $L \approx d/2$ ($L \approx r$) for a hinged edge (HE) and $L \approx d/3$ ($L \approx 2\,r/3$) for a clamped edge (CE) is given by simple analysis of the expected deformed shape of the square (circular) plate or shell.

For the square plate it is then the surface normal load q^3 with intensity $q = \mathcal{M}(q^3) \approx E\,h^4/L^4$ (i.e. $q \approx 16\,E\,h^4/d^4$ in case HE and $q \approx 81\,E\,h^4/d^4$ in case CE) which according to the asymptotic analysis causes the deflection u_3 of the thickness magnitude $\left(\mathcal{M}(u_3) \approx h\right)$ and requires the use of the geometrically nonlinear bending theory as the lowest order approximation.

This is actually confirmed by calculation, e.g. in [10] we find

$$\begin{aligned} HE:&\quad \max|u_3| = 0.753\,h \quad \text{for} \quad q = \frac{19\,E\,h^4}{d^4},\\ CE:&\quad \max|u_3| = 0.909\,h \quad \text{for} \quad q = \frac{96\cdot 1\,E\,h^4}{d^4}. \end{aligned} \tag{5.1}$$

When $q \ll E\,h^4/L^4$ then according to (4.1.3) the zeroth order approximation is represented by the linear bending theory and from (4.5), (4.18) we deduce $\mathcal{M}(u_3) \approx \mathcal{P}\,L^4/(E\,h^3)$ and $\mathcal{M}(s^{\alpha\beta}) \approx \mathcal{P}\,L^2/h^2$, which values are in good agreement with classical results of [11]

$$\begin{aligned} HE:&\quad \mathcal{M}(u_3) \approx 0.0625\,\frac{q\,d^4}{E\,h^3}, \qquad \mathcal{M}(s^{\alpha\beta}) \approx 0.25\,\frac{q\,d^2}{h^2},\\ &\quad [11] - \max|u_3| = 0.0487\,(1-\nu^2)\,\frac{q\,d^4}{E\,h^3}, \qquad \max|s^{11}| = 0.2874\,\frac{q\,d^2}{h^2};\\ CE:&\quad \mathcal{M}(u_3) \approx 0.0125\,\frac{q\,d^4}{E\,h^3}, \qquad \mathcal{M}(s^{\alpha\beta}) \approx 0.111\,\frac{q\,d^2}{h^2},\\ &\quad [11] - \max|u_3| = 0.01512\,(1-\nu^2)\,\frac{q\,d^4}{E\,h^3}, \qquad s_i^{11} = 0.1386\,\frac{q\,d^2}{h^2}, \end{aligned} \tag{5.2}$$

where ν is POISSON's ratio and s_i^{11} is the value of $|s^{11}|$ in the centre of the plate.

It is to be noted once more that all comparisons are being made in the *order of magnitude* sense.

In case of the clamped circular plate with $q \gg E\,h^4/L^4$ $\left(= 81\,E\,h^4/(16\,r^4)\right)$ we deduce the following information and compare it with classical results from [11]

$$\begin{aligned} &\mathcal{M}(u_3) \approx 0.592\,r\left(\frac{q\,r}{E\,h}\right)^{1/3}, \qquad \mathcal{M}(s^{\alpha\beta}) \approx 0.763\,q\left(\frac{E\,r^2}{q\,h^2}\right)^{1/3},\\ &[11]:\ \max|u_3| = 0.622\,r\left(\frac{q\,r}{E\,h}\right)^{1/3}, \qquad \max|s^{11}| = 0.423\,q\left(\frac{E\,r^2}{q\,h^2}\right)^{1/3}. \end{aligned} \tag{5.3}$$

Considering this case with edge compressive uniform load $t^1 = -\,t$, expression (4.17) gives the critical value of this load by

$$t_{\text{crit}} = \frac{E\,h^2}{r^2} \qquad \left([11]:\ t_{\text{crit}} = \frac{(3.83)^2\,E\,h^2}{12\,(1-\nu^2)\,r^2}\right), \tag{5.4}$$

and we see once more that the simple calculation based on the order of magnitude relations of the method of asymptotic integration provides rational first information on the stresses and displacements.

Passing to the shell examples, we consider the problem of finding the critical value of the uniform surface load $q^3 = -\,q$ for spherical shell with middle surface radius R_S. Expressions (4.17) indicate that here the characteristic wave length L plays a decisive role. For non-shallow shells with no local buckling effects, the case (4.17.1) can be expected, yielding the critical value

$$q_{\text{crit}} \approx \frac{E\,h^2}{R_S^2} \qquad \left([11]:\ q_{\text{crit}} = \frac{2\,E\,h^2}{R_S^2\,3\,\sqrt{(1-\nu^2)}}\right), \tag{5.5}$$

which actually is of the same order of magnitude as the classical value [11]. When the tendency to local buckling occurs, then L decreases and also the case (4.17.2) can take place with the critical value smaller than (5.5), which case is often reported on the basis of the test results [12]. The factors which can influence the value of L and thus also the order of magnitude of the critical load are imperfections of any sort and also the type of the edge conditions, as actually noticed in [12]. Other instructive shell examples are given in [13], where some predictions made on basis of the asymptotic analysis are compared with numerical results for two spherical caps.

When summarising the advantages of the method of asymptotic integration, the results presented in this paper show that this method provides us with a procedure allowing the approximate solution of the shell problem to be constructed with the asymptotic accuracy required, gives us the possibility to obtain the shell and plates problems classification in terms of the shell data and to deduce the basic information on the shell behaviour prior to the numerical solution of the differential equations of the proper classical shell theory.

References

1 GOLDENVEJZER, A. L., Derivation of an approximate theory of shells by means of asymptotic integration of the equations of the theory of elasticity (in Russian). Prikl. Math. Mekh. **27**, pp. 593–608 (1963).
2 GREEN, A. E., On the linear theory of thin elastic shells. Proc. R. Soc. **266**, pp. 143–160 (1962).
3 GREEN, A. E. and NAGHDI, P. M., Some remarks on the linear theory of shells. Q. J. Mech. appl. Math. **18**, pp. 257–276 (1965).
4 RUTTEN, A. S., Asymptotic Approximation in the Threedimensional Theory of Thin and Thick Elastic Shells. Doctoral Dissertation, Delft, 1971.
5 NOVOTNÝ, B., On the asymptotic integration of the threedimensional non-linear equations of thin elastic shells and plates, Int. J. Solids Structures **6**, pp. 433–451 (1970).
6 NOVOTNÝ, B., The method of asymptotic integration in the non-linear theory of shells (in Slovak) Report of Slovak Academy of Sciences, Bratislava, 1970.
7 GUSEYN-ZADE, M. I., On necessary and sufficient conditions of existence of a damped solutions of the plane problem of the theory of elasticity for semi-infinite strip (in Russian), Prikl. Math. Mekh. **29**, pp. 393–399 (1965).
8 NOVOTNÝ, B., On existence conditions of a damped solution of the plane problem for orthotropic semi-infinite strip (in Slovak), Stavebnícky časopis **16**, pp. 467–476 (1968).
9 SRUBSCHIK, L. S., On asymptotic integration of the systems of nonlinear equations of plate theory (in Russian), Prikl. Math. Mekh., **28**, pp. 335–349 (1964).
10 KORNISHIN, M. S., ISANBAYEVA, F. S., Flexible Plates and Panels (in Russian), Nauka, Moscow, 1968.
11 TIMOSHENKO, S., WOINOWSKY-KRIEGER, S., Theory of Plates and Shells, McGraw-Hill, New York–Toronto–London, 1959.
12 WANG, L. R. L., RODRIGUEZ-AGRAIT, L., LITLE, W. A., Effect of boundary conditions on shell buckling, J. Engn. Mech. Div. ASCE **92**, No. EM6, pp. 101–116 (1966).
13 NOVOTNÝ, B., The use of results of method of asymptotic integration for classification of twodimensional classical shell theories (in Slovak), Stavebnícky časopis **21**, pp. 348–356 (1973).

Eingereicht am 5. 12. 1972

Anschrift: Ing. B. NOVOTNÝ CSc, Institute of Construction and Architecture of the Slovak Academy of Sciences, Bratislava, Dubravská cesta 4, ČSSR

ZAMM 54, 157 – 164 (1974)

S. Tsugé / T. Fujiwara

On the Propagation Velocity of a Detonation-Shock Combined Wave

Für die freie Detonation eines Gases in einer inaktiven Umgebung werden eine quasi-eindimensionale Analysis und eine verallgemeinerte Chapman-Jouget-Bedingung entwickelt. Aus den Ergebnissen werden zwei charakteristische Merkmale abgeleitet: Für die Verbrennungswärme, den Ladungsdurchmesser und das Molekulargewicht des umgebenden Gases existiert je eine untere Grenze, und für eine Parameterkombination oberhalb eines kritischen Wertes werden zwei Ausbreitungsgeschwindigkeiten gefunden. Daher ist das Ausbleiben der Detonation einer freien Gassäule eher ein hydrodynamisches als ein chemisches Problem.

A quasi-onedimensional analysis and a generalized Chapman-Jouguet condition are developed for the gaseous free detonation in an inert environment. Two characteristic features derived from the results are that there exists a lower limit for each of the heat of combustion, the charge diameter and the molecular weight of the environmental gas and that two propagation velocities are for an above-critical combination of the parameters. Therefore the detonation failure of a free gas column is a hydrodynamic rather than a chemical phenomenon.

Для свободного взрыва газа в инертной окружающей среде излагаются квази-одномерный анализ и обобщённые критериальные условия Шепмэна-Жюгета. Из результатов извлекаются две характерные приметы: для каждой теплоты горения, диаметра заряда и молекулярного веса окружающего газа существует нижний предел, а для параметров превышающих критическое сочетание возможны две скорости распространения. Следовательно неудачные взрывы свободного газового столба являются скорее гидродинамическим чем химическим феноменом.

1. Introduction

There have been many investigations [1]—[6] on the "free" detonations, i.e., those detonations which propagate in unconfined solid explosives or in combustible gas column separated from the environment inert gas by thin films. The main interest lies in predicting the velocity deficit ΔD,

$$\Delta D = (D_{C-J})_{\text{plane}} - D\,, \tag{1.1}$$

and in accounting for the occurrence of detonation failure. In Eq. (1.1) $(D_{C-J})_{\text{plane}}$ is the propagation velocity of a plane Chapman-Jouguet detonation and D is the actual propagation velocity of the free detonation.

A thorough examination of the existing theories shows that they are subject to either of the following defects:

a) ΔD is assumed as small; the theories are exact only to the first order in ΔD [7]. This eliminates the possibility of explaining the occurence of detonation failure, and/or

b) the Chapman-Jouguet condition is not correctly posed; in earlier papers it does not play the role of the eigen-value condition which rules out the inappropriate solution, but it is borrowed incorrectly from the theory of the plane wave.

The present investigation aims at constructing the theory which is free from these faults and at obtaining the clearcut picture of the detonability limit by confining ourselves to the case where the following model is realistic (see Fig. 1):

(i) The detonated gas flow is quasi-onedimensional and preceded by $Z-N-D$ detonation front.

(ii) The motion of the surrounding gas is regarded as Newtonian [8] due to its hypersonic character.

(iii) The explosive is an ideal gas and obeys a one-step combustion reaction.

It is evident that the velocity deficit ΔD arises from the additional wave drag due to the shock wave induced in an inert gas. To maintain the steady propagation the explosive has to be powerful enough to overcome the "wave resistance". It is natural, therefore, to expect that there would exist a lower limit for the exothermicity and the charge diameter below which no steady propagation of the detonation-shock combined wave is possible. It is also expected that a very soft (low density) surrounding inert gas which absorbs the major fraction of the detonation energy will bring about the detonation failure. These phenomena have been observed by several experimenters [3], [9], [10].

The objective of the following analysis is to clarify the dependence of the velocity deficit on these factors:

$$\frac{\Delta D}{D_{C-J}} = F\left(\frac{Q}{C_p T_0}, \frac{l}{d}, \frac{\varrho_{00}}{\varrho_0}\right), \tag{1.2}$$

where $C_p T_0$ is the thermal enthalpy of the unburnt explosive, l the thickness of the detonation wave, d the charge diameter, and ϱ_{00} and ϱ_0 the densities of the inert gas and the unburnt explosive, respectively. The dependence of the expression (1.2) on d implies that the detonation velocity is no longer a physico-chemical constant, but is a quantity obtainable only through the complete solution of the equations of the fluid-dynamic field. The present theory was applied, to obtain the same dependence, to oxyhydrogen mixtures [11] and also to oxyozone mixtures to check the effect of the external pressure [12], after the present paper was written (the main results are included in Ref. [13] in Japanese).

So long as the outer flow is concerned, the present problem is apparently similar to that of the hypersonic flow round a (deformable) obstacle (see Fig. 1). The only but intrinsic difference is that in the latter we have only to solve the problem as a forced motion in which the free stream velocity is *prescribed*, while in the former the value D should be determined self-consistently as a kind of *eigen-value*. Hence the problem requires an additional condition besides three conservation equations, and it is this condition that play a central role in the problem of detonations. For a plane detonation it is identified as the $C-J$ condition, and for axially-symmetric detonation of an explosive rod, on the other hand, it results in the WOOD-KIRKWOOD condition [4]:

$$\frac{\partial v}{\partial r} = -\left(\frac{\varrho\,\beta}{2\,C_p}\right)\left(\frac{\partial e}{\partial \Lambda}\right)_{p,\varrho} \mathfrak{M} \qquad \text{at the sonic point:} \qquad M = 1\,, \tag{1.3}$$

where β is defined as $(\partial\varrho^{-1}/\partial T)_{p,\Lambda}$, e the specific internal energy, Λ the progress variable and $\mathfrak{M}$ the mass production rate of the final product per unit mass.

In principle, this eigen-value condition, or the generalized CHAPMAN-JOUGUET condition, has a function such that it singles out a physically acceptable solution which is free from the point of infinite derivative anywhere in the flow field. Otherwise, we should have a series of solutions all but one of which exhibit infinite acceleration at the sonic pount. Actually this condition should be *derived* through a standard mathematical procedure on the singular point and should not be given apriori. Therefore the actual form of this conditon differs for the different choice of model or of the governing equation. In most of the earlier papers except that of WOOD and KIRKWOOD [4], the generalized $C-J$ condition is given not from the above-mentioned standpoint but from an analogy to the plane problem.

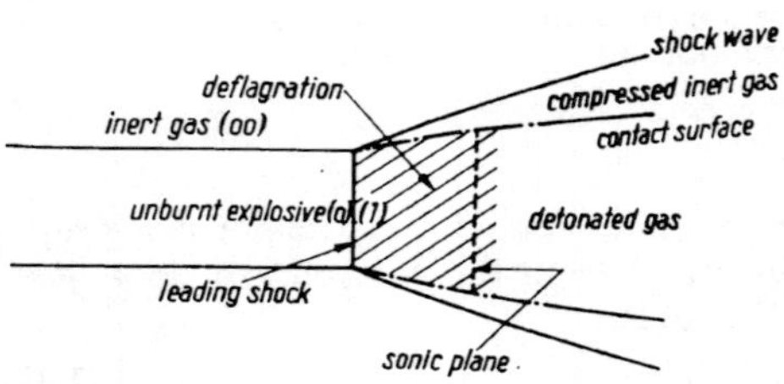

Fig. 1. Sketch of a free detonation wave propagating in a flexible tube

In this paper we adopt the quasi-onedimensional model for the reactive flow (Section 2) and carry out the above procedure of obtaining the generalized $C-J$ condition (Section 3). The last two sections (5 and 6) are devoted to the actual calculation for the ozone decomposition detonation and the existence of double-valued propagation velocities and of the detonability limit is exhibited.

Throughout the calculation we adopt the $Z-N-D$ model, i.e., we assume that the regions (0) and (1) in Fig. 1 are separated by a shock wave with no chemical reaction, and the shocked state (1) is characterized by

$$\Lambda = 0\,, \tag{1.4}$$

$$\frac{p_1}{p_0} = r_p(M_0)\,, \qquad \frac{T_1}{T_0} = r_T(M_0)\,, \tag{1.5}$$

where $M_0 = D/a_0$ is the detonation Mach number, and r_p and r_T are the pressure and the temperature ratios across the shock wave, respectively, given as

$$r_p = (2\,\gamma\,M_0^2 - \gamma + 1)/(\gamma + 1)\,, \qquad r_T = r_p(M_0)\,(2\,M_0^{-2} + \gamma - 1)/(\gamma + 1)\,. \tag{1.6}$$

The symbols used in this paper are as follows:

- A — σ/σ_1, area ratio
- a — velocity of sound
- C_p — specific heat at constant pressure
- D — detonation velocity
- ΔD — $(D_{C-J})_{\text{plane}} - D$, velocity deficit
- d — charge diameter
- E — 24150 cal/mole, activation energy of ozone decomposition reaction
- k — ϱ_{00}/ϱ_0, density ratio between environment and unburnt explosive
- k_1 — reaction rate constant
- k_2 — reaction rate constant
- l — characteristic reaction length
- M — local flow MACH number
- M_0 — D/a_0, detonation MACH number
- P — $(p/p_1)\,A$, dimensionless pressure
- p — pressure
- Q — heat of reaction
- q — $Q/C_p\,T_0$, dimensionless heat of reaction
- R — universal gas constant
- r_p — pressure ratio across shock wave
- r_T — temperature ratio across shock wave
- T — temperature
- u — flow velocity
- V — $(\varrho_1/\varrho)/A$, dimensionless specific volume
- W — mass production rate of O_2
- W_i — molecular weight of i-th species
- X — third body in chemical reaction
- X_i — mole concentration of i-th species in per cent
- x — coordinate
- Y_i — mass fraction of i-th species
- α — frequency factor in ozone decomposition reaction
- γ — specific heat ratio
- Λ — progress variable of reaction
- λ — d/l, dimensionless charge diameter
- ϱ — density
- σ — cross sectional area of charge

subscripts:

- 0 — initial state of explosive charge
- 00 — initial state of environment
- 1 — immediately behind leading shock wave, or NEUMANN state
- O_2 — quantity of molecular oxygen
- O_3 — quantity of ozone
- [ ] — molar concentration

2. Quasi-Onedimensional Flow with Deflagration

Strictly speaking, the deflagration region should be determined from three conservation laws in the form of partial differential equations describing the axially symmetric reacting gas flow. This, however, is a difficult task, for the flow is unavoidably rotational and nonisentropic, and also the finite rate of reaction has to be taken into account. The difficulty would be eliminated if we can describe the gasdynamic field through a system of *ordinary* differential equations by adopting a quasi-onedimensional model. In principle such a treatment is admissible only when the cross-sectional area is slowly varying in the stream direction.

The fundamental equations of the quasi-onedimensional steady flow with variable cross sectional area $\sigma(x)$ are written in the form

$$\varrho_1 u_1 \sigma_1 = \varrho u \sigma , \tag{2.1}$$

$$\mathrm{d}\,[(\varrho u^2 + p)\,\sigma] = p\,\mathrm{d}\sigma , \tag{2.2}$$

or

$$(\varrho_1 u_1^2 + p_1)\,\sigma_1 = (\varrho u^2 + p)\,\sigma - \mathcal{D} , \tag{2.2'}$$

with $\mathcal{D} = \int_{\sigma_1}^{\sigma} p\,\mathrm{d}\sigma$,

$$\frac{1}{2} u_1^2 + c_p T_1 = \frac{1}{2} u^2 + c_p T - Q \Lambda , \tag{2.3}$$

and

$$\varrho u \frac{\mathrm{d}\Lambda}{\mathrm{d}x} = W . \tag{2.4}$$

In the above expressions the subscript 1 refers to the state immediately behind the leading shock wave. The coordinate employed is a shock one with its origin at position of the leading shock and with its positive direction toward the deflagration region. $\mathcal{D}$ is the drag acting on the sidewall portion of duct-shaped body from $x = 0$ to the location where the unsubscribed quantities prevail.

For later convenience we introduce the following dimensionless variables:

$$A = \sigma/\sigma_1 , \qquad V = (\varrho_1/\varrho)/A , \qquad P = (p/p_1)\,A . \tag{2.5}$$

Eq. (2.2) is rewritten in terms of these variables as

$$\frac{\mathrm{d}}{\mathrm{d}\,(\ln A)} \left(V + \frac{P}{\gamma M_1^2} \right) = \frac{P}{\gamma M_1^2}, \tag{2.6}$$

where M_1 is the MACH number immediately behind the leading shock wave, connected with M_0 by

$$M_1^2 = M_0^2\, r_T / r_p^2 . \tag{2.7}$$

In the same way we have from Eqs. (2.1) and (2.3)

$$V = - \frac{P}{(\gamma - 1)\, M_1^2} + \left[\left\{ \frac{P}{(\gamma - 1)\, M_1^2} \right\}^2 + \frac{2}{(\gamma - 1)\, M_1^2} \left(1 + \frac{q\,\Lambda}{r_T} \right) + 1 \right]^{1/2}, \tag{2.8}$$

where q denotes the dimensionless heat of combustion:

$$q = Q/(c_p T_0) . \tag{2.9}$$

Eqs. (2.4), (2.6) and (2.8) constitute a system of equations describing a nozzle flow with a chemical reaction and with variable cross-sectional area. We have, however, a freely deformable tube wall as noted previously. The single condition prevailing there is continuity of the pressure, i.e., $A(x)$ is not prescribed but has to be solved consistently with the outer gas flow.

3. Generalized Chapman-Jouguet Condition

Before carrying out the solution procedure we have to show the key for the unique determination of M_0 appearing in the equation as a free parameter. Elimination of Λ and V from Eqs. (2.4), (2.6) and (2.8) yields the following equation:

$$\frac{\mathrm{d}P}{\mathrm{d}x} = \frac{1 + (\gamma - 1)\, M^2}{1 - M^2} \left[\frac{P}{A} \frac{\mathrm{d}A}{\mathrm{d}x} - \frac{A}{P} Z(M)\, q\, W \right], \tag{3.1}$$

where we have put

$$Z(M) = \frac{\gamma M_0}{r_p^2\, \varrho_0\, a_0 \{1 + (\gamma - 1)\, M^2\}} . \tag{3.2}$$

In the above expression M is the local MACH number which is related to V and P by

$$M^2 = M_1^2\, V/P . \tag{3.3}$$

Eq. (3.1) shows the influence of the chemical reaction on the relation between the pressure and the cross-sectional area. It degenerates into the well-known "LAVAL nozzle" equation if chemical reaction is absent. Exothermicity $q > 0$ makes the numerator of r.h.s. of Eq. (3.1) decrease which reflects that the heat release has a hydrodynamical effect equivalent to contracting the stream tube. It is this effect that enables the shocked gas (1) to accelerate beyond the sonic point in a monotonically diverging stream tube.

At the sonic point, however, the pressure gradient goes to infinity unless the numerator of Eq. (3.1) vanishes simultaneously. This means that to have a physically plausible solution the following condition is to be fulfilled:

$$\frac{\mathrm{d}A}{\mathrm{d}x} = \left(\frac{A}{P}\right)^2 Z(M)\, q\, W \qquad \text{at} \qquad M^2 = 1\,. \tag{3.4}$$

This condition allows us to determine the free parameter M_0, whence it is referred to as a generalized CHAPMAN-JOUGUET condition (c.f. the WOOD-KIRKWOOD condition (1.3) for the axially symmetric configuration). For a plane detonation (A = constant) this condition leads to the well-known form: $W = 0$ at $M = 1$. The condition (3.4) imposes a specific relation on the chemical reaction rate and the stream tube divergence in passing a sonic point and it demonstrates that the sonic point is located at the position where the combustion is not yet completed. In this sense the detonation of this type is pathological.

4. The Newtonian Approximation for Outer Flow

Eqs. (2.4), (2.6) and (2.8) are still insufficient to determine the unknowns V, P, A and Λ. Evidently the lacking equation is the one describing the variation of cross-sectional area. In contrast with the flow in a LAVAL nozzle the only condition available for the determination of $A(x)$ is the continuity of pressure across the tangential discontinuity. Thus the outer flow of the inviscid gas should be solved.

In principle this is the problem of the gasdynamic flow round a body of revolution. Due to the hypersonic character and the monotonical increase of A with x, the NEWTON model [8] is expected to give a valid approximation:

$$p = p_0 + \varrho_{00}\, D^2 \left(\frac{\mathrm{d}\sigma}{\mathrm{d}x}\right)^2 \Big/ \left[\left(\frac{\mathrm{d}\sigma}{\mathrm{d}x}\right)^2 + 4\,\pi\,\sigma\right]$$

or, alternatively,

$$\frac{\mathrm{d}A}{\mathrm{d}x} = \left(\frac{4\,\pi}{\sigma_0}\right)^{1/2} \left[\frac{A\,(P\,r_p - A)}{(1 + \gamma\,k\,M_0^2)\,A - r_p\,P}\right]^{1/2}, \tag{4.1}$$

where σ_0 is the initial cross-sectional area of the explosive and

$$k = \varrho_{00}/\varrho_0 \tag{4.2}$$

is the density ratio between the environment and the unreacted explosive. Models other than the NEWTONian, e.g., the "tangent cone" model etc. would also be available, but we proceed in the following on the NEWTON model because of its simplicity.

5. The Solution Procedure

By eliminating the variables x and V from Eqs. (2.4), (3.1) and (4.1) we lead to the following system of ordinary differential equations:

$$\frac{\mathrm{d}P}{\mathrm{d}\Lambda} = (1 - M^2)^{-1}\, f(P, A, \Lambda; M_0)\,, \qquad \frac{\mathrm{d}A}{\mathrm{d}\Lambda} = g(P, A, \Lambda; M_0)\,, \tag{5.1}$$

where f and g are defined, respectively, by

$$f = [1 + (\gamma - 1)\, M^2]\frac{P}{A}\, g - \frac{\gamma\, M_0^2\, q}{r_p^2\, P}, \qquad g = \frac{4\, M_0}{\lambda\, \overline{W}} \left\{\frac{r_p\, P - A}{A\,[(1 + \gamma\, k\, M_0^2)\, A - r_p\, P]}\right\}^{1/2}. \tag{5.2}$$

In the above expression $\overline{W}$ is the dimensionless reaction rate defined by $\overline{W} = (l/\varrho_0\, a_0)\, W$ and λ is the following dimensionless tube diameter:

$$\lambda = d/l\,. \tag{5.3}$$

We are seeking for the solution which satisfies the initial condition

$$A = P = 1 \qquad \text{at} \qquad \Lambda = 0\,, \tag{5.4}$$

and the eigen-value condition

$$g = g^* = \frac{M_0^2\, A^*\, q}{P_*^2\, r_p^2} \qquad \text{at} \qquad M = 1\,, \tag{5.5}$$

where the asterisk refers to the quantities at the sonic point. Assuming the rate function W in the form

$$W = \omega\, \varrho^n (1 - \Lambda)\, e^{-E/(RT)} , \tag{5.6}$$

the singularity at the sonic point can be shown to behave as a saddle one in contrast with the fact that the $C - J$ point for the plane detonation has a nodal one [15].

The actual numerical calculation was carried out for the oxygen-diluted ozone as an explosive gas and the argon/oxygen as an environmental gas. Ozone was chosen simply because its reaction mechanism is well understood and because its apparent one-step character simplifies the calculation remarkably. It is assumed here that the ozone decomposition reaction obeys the following scheme [15]:

$$O_3 + X \overset{k_1}{\rightarrow} O_2 + O + X , \qquad O + O_3 \overset{k_2}{\rightarrow} O_2 + O_2 , \tag{5.7}$$

where [16]

$$k_1 = 4.61 \cdot 10^{15} \exp\left(-24150/(RT)\right) \quad \mathrm{cm^3/mol \cdot sec} ,$$

$$k_2 = 3 \cdot 10^{13} \exp\left(-6000/(RT)\right) \quad \mathrm{cm^3/mol \cdot sec} .$$

Adopting the steady-state assumption for the atomic oxygen the reaction rate W for the product gas O_2 is written in the form

$$W = \omega\, \varrho^2 (1 - \Lambda)\, e^{-E/(RT)} , \tag{5.8}$$

$$\omega = 3\,\alpha (W_{O_2}/W_{O_3}^2) \left[\frac{3}{2} - \frac{1}{2}(1 - \Lambda)\,(Y_{O_3})_{\text{initial}}\right],$$

$$E = 24150 \text{ cal/mole} , \qquad \Lambda = 1 - Y_{O_3}/(Y_{O_3})_{\text{initial}} ,$$

where Y_{O_3} is the mass fraction of ozone. W_{O_2} and W_{O_3} express the molecular weight of oxygen and ozone, respectively.

The parameters and the physical quantities of the explosive gas are shown in Table 1, where the following numerical values have been adopted:

$$p_0 = 1 \text{ atm} , \qquad T_0 = 300\ {}^\circ\mathrm{K} , \qquad Q = 34535 \text{ cal/mole} ,$$

$$(c_p)_{O_2} = 7.0 \text{ cal/mole} \cdot {}^\circ\mathrm{K} , \qquad (c_p)_{O_3} = 9.0 \text{ cal/mole} \cdot {}^\circ\mathrm{K} .$$

Table 1. Parameters and physical quantities for the plane CHAPMAN-JOUGUET detonation of oxyozone mixtures

Ozone concentration $(X_{O_3})_{\text{initial}}$ (mole%)	$(M_{C-J})_{\text{plane}}$	p_1/p_0	T_1/T_0	Thickness of Plane $C-J$ Detonation $(l_{C-J})_{\text{plane}}$ (cm)	Sound Speed a_0 (m/sec)
7	2.698	8.33	2.341	$1.073 \cdot 10^{0}$	324.6
8	2.837	9.23	2.492	$3.637 \cdot 10^{-1}$	323.8
9	2.969	10.12	2.642	$1.413 \cdot 10^{-1}$	323.1
10	3.094	11.00	2.791	$6.133 \cdot 10^{-2}$	322.3
11	3.213	11.88	2.938	$2.919 \cdot 10^{-2}$	321.5
12	3.327	12.75	3.084	$1.500 \cdot 10^{-2}$	320.8
13	3.437	13.61	3.229	$8.232 \cdot 10^{-3}$	320.0
14	3.542	14.47	3.373	$4.778 \cdot 10^{-3}$	319.3
15	3.644	15.33	3.516	$2.911 \cdot 10^{-3}$	318.5
16	3.743	16.17	3.658	$1.850 \cdot 10^{-3}$	317.8
17	3.838	17.02	3.799	$1.219 \cdot 10^{-3}$	317.1
20	4.108	19.52	4.217	$4.188 \cdot 10^{-4}$	314.9

The characteristic reaction length l is defined as

$$l = \frac{a_0 (W_{O_3})^2}{2\, W_{O_2}\, \varrho_0\, \alpha} \exp\left(\frac{E}{RT_1}\right), \qquad \alpha = 4.61 \cdot 10^{15} . \tag{5.9}$$

The numerical calculation was carried out along the following steps:

(i) Assume M_0 $\left(< (M_{C-J})_{\text{plane}}\right)$,

(ii) carry out the integration of Eq. (5.1) under the initial condition (5.4),

(iii) check the point at which the numerator or the denominator of $\mathrm{d}P/\mathrm{d}\Lambda$ changes sign. When it is expected that the denominator changes sign first, the integration must be done interchanging the dependent and independent variables.

(iv) Search the numerical value of M_0 for which the two zeros coincide. This procedure cannot be accomplished, in principle, only through the numerical calculation. However, as shown in the next section, the crossing point of the loci of each zero can be predicted within sufficient accuracy, since very slight amount of variation of assumed M_0 affects whether we get the zero of the numerator or of the denominator first. Thus the eigen-value M_0 can be obtained with comparative ease. The HITAC 5020 electronic computer of Tokyo University was used.

6. Velocity Deficit and Detonability Limit

A typical numerical procedure is shown in Fig. 2. In Fig. 2.a) where initial ozone concentration $(X_{O_3})_{\text{initial}} = 10$ mole %, charge diameter $d = 10$ cm, and argon is the environment, the locus of the sonic point, i.e., the locus of the zero of the denominator of $dP/d\Lambda$ intersects with that of f twice, whereas for a slightly less exothermic case $\left((X_{O_3})_{\text{initial}} = 7 \text{ mole } \%\right)$ there exists no intersection for any value of M_0. Thus there is no solution for such a less exothermic case (Fig. 2. b)).

In more detail one meets the following four cases corresponding to specified sets of parameters k, λ and M_0:

(I) Neither the sonic point nor the $f = 0$ point appears.
(II) Only the $f = 0$ point appears.
(III) First the $f = 0$ point and next the sonic point appears.
(IV) The $f = 0$ point appears after the sonic point is passed.

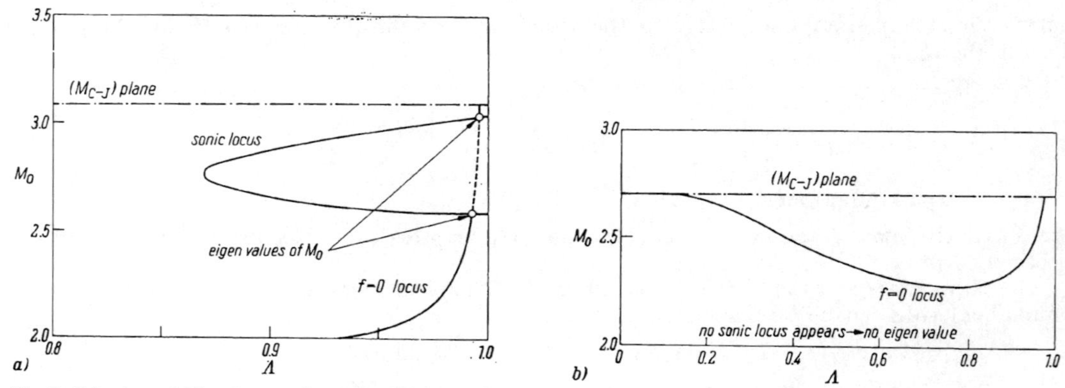

Fig. 2. Behaviour of $M_0 - \Lambda$ curves for various M_0 assumed

(a) Above detonation limit: For such combinations of charge diameter (d = 10 cm), initial ozone concentration (X_{O_3} = 10%) and environment gas (= Ar), Eq. (5.1) has two eigen solutions satisfying $C-J$ condition (5.5)

(b) Below detonation limit: For such combinations of d (= 10 cm), X_{O_3} (= 7%) and k (corresponding to Ar), no solution satisfying $C-J$ condition (5.5) can be found

The numerical procedure in the case (I) or (II) meets no trouble. For the case (III) one meets the sonic point after having passed the $f = 0$ point. As mentioned in the previous section such a point can be passed over by exchanging the role of dependent and independent variables. The case (IV) can be treated similarly. In the present calculation, however, only the direct integration of Eq. (5.1) up to the sonic location is carried out. The dotted curve in Fig. 2.a) is not calculated actually. It should be noted that such treatment is quite sufficient to obtain the two cross points.

Obtained MACH number deficit for different combinations of $(X_{O_3})_{\text{initial}}$, d and environmental gases is shown in Fig. 3. Two characteristic features obtained for solid detonations by EYRING et al. [2] and for confined gaseous detonations by ZELDOVICH et al. [22] are exhibited.

(I) There exist two propagation velocities for a specified set of the charge diameter, the heat of combustion and the molecular weight of the inert environment.
(II) There is no solution if the heat of combustion falls short of a critical value.

Of the two propagation velocities the one with smaller velocity deficit is confluent with the plane CHAPMAN-JOUGUET value with increasing d or k. Concerning the other branch nothing conclusive can be said at the present stage. As to the calculated results we can draw some conspicuous features on the slower branch.

a) It is pathological in the sense that at most about 2% of the reactant remains unburnt at the sonic point.
b) Reflecting the weakness of the igniting shock wave, it has by far the thicker reaction zone than the faster branch.

On the basis of these facts the slower branch is explained as follows: Since a supersonic flow is realized beyond the sonic point, any events taking place at its downstream has no influence on the upstream phenomena, e.g., on M_0. Therefore the fuel burning downstream of the sonic point is wasted, and the propagation velocity is reduced. On the other hand, thickened reaction zone increases the wave drag $\mathcal{D}$ in Eq. (2.2').

Double-valued character of the detonation velocity has been observed for gas detonations [17], [18] in tubes with specific wall properties and for the detonation of uncased solid explosives [19]. It should be stressed that the obtained double-valuedness differs from the former in its nature. For, in the experiments of MUNDAY, UBBELOHDE and WOOD, it is clarified that the origin of double-valuedness is the chemical kinetics. It is also observed that most explosive remains unburnt after the passage of low velocity detonations in solids.

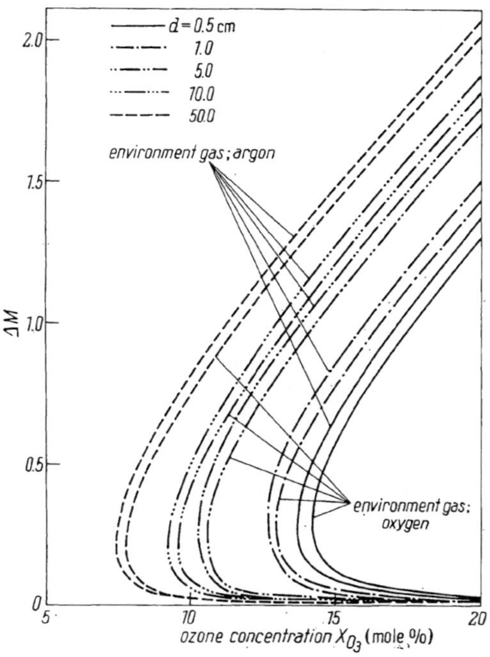

Fig. 3. Deficit ΔM of propagation MACH number from that of plane CHAPMAN-JOUGUET detonation in O_3-O_2 mixture, with initial ozone concentration varied

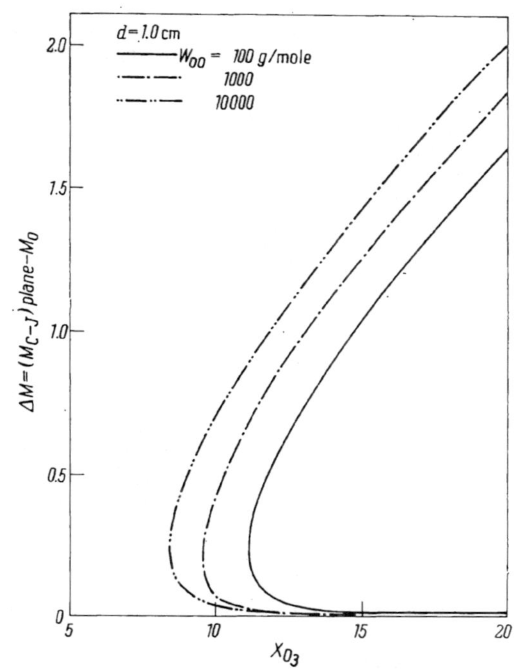

Fig. 4. Initial concentration of O_3 v.s. the MACH number deficit for hypothetically heavy gases as the environment

In the present calculation the detonation limit is obtained in an unambiguous manner without introducing any supressing reaction in chemical kinetics (such as HO_2 formation in hydrogen-oxygen reaction). This means that so far as free detonations are concerned the existence of detonability limit is of hydrodynamical rather than of chemical nature [20].

We have to remark on the validity of the above results. The second branch and the merging point of two branches fall in the region where M_0 is not large enough to permit the use of the NEWTONian approximation. To show that the double-valuedness and the detonability limit are brought about not through the NEWTONian approximation, we repeat the above calculation for environmental gases with hypothetically high molecular weight. Since the MACH number is proportional to the square root of molecular weight, such examples provide a high propagation MACH number with respect to the environment even near the detonability limit and on the second branch. Results for such cases are shown in Fig. 4 which confirms the existence of detonability limit and double-valuedness.

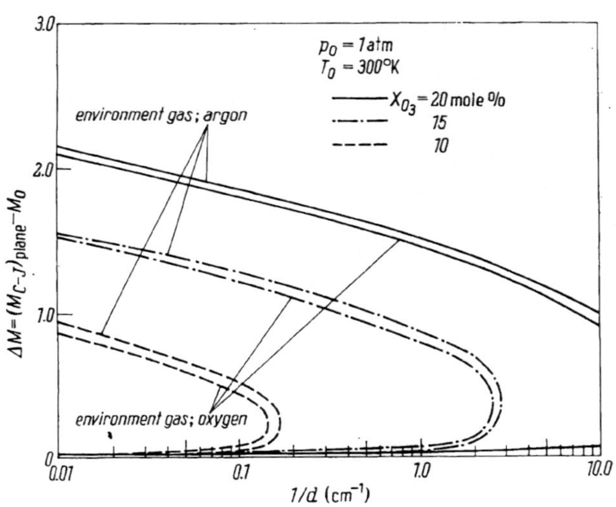

Fig. 5. Deficit ΔM of propagation MACH number from that of plane CHAPMAN-JOUGUET detonation versus reciprocal charge diameter in $O_3 - O_2$ mixture with argon/oxygen as environment gases

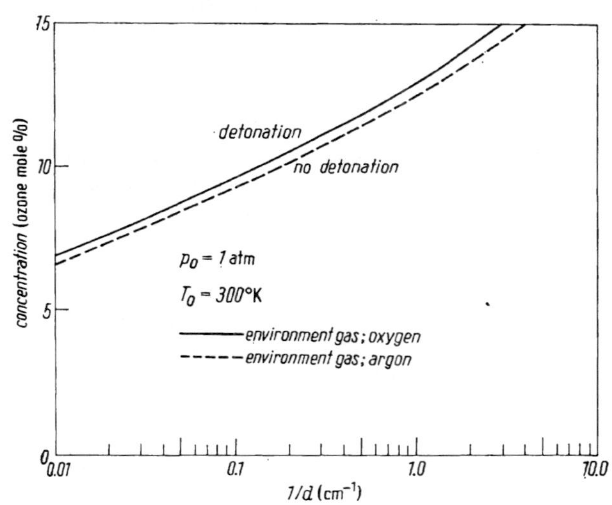

Fig. 6. The detonability limit of O_3-O_2 mixture with oxygen and argon as environmental gases

A few remarks are necessary on the choice of the environmental gas. From (5.2) the requirement that denominator of g be positive at $\Lambda = 0$ yields the condition

$$k > 2/(\gamma + 1) .$$

When the unreacted explosive and the inert environment are in thermodynamical and mechanical equilibrium (having the same temperature and pressure), the above condition is written as

$$W_{00}/W_0 > 2/(\gamma + 1) . \tag{6.2}$$

This precludes the light gases (such as the helium or the hydrogen) as the environment. Though it is not certain whether the use of a light gas leads to the detonation failure or to the detonation propagation of a different type, it has been confirmed experimentally [3] that the flow configuration sketched in Fig. 1 is no longer observed under such circumstances.

By rearranging the foregoing results we have a figure which shows the velocity deficit as a function of reciprocal of the charge diameter (Fig. 5). The two features mentioned above are kept unchanged and the critical diameter below which no steady detonation exists is defined clearly.

Fig. 6 shows the region where the steady solution exists in the plane of the concentration and the reciprocal diameter. In agreement with what was expected in Section 1, a heavier environmental gas allows a wider range of detonability. It is of some interest to note that for the range of realistic dimension of experimental apparatus the critical concentration is located near $(X_{O_3})_{\text{initial}} = 10$ mole % which is the experimentally observed critical value for confined plane detonations [21].

Acknowledgements

The authors wish to express their thanks to Dr. HARUO KAWADA, Tokyo Institute of Technology, for his kind advices.

References

1 JONES, H., Proc. Roy. Soc. (London) **A189**, 415 (1947).
2 EYRING, H., POWELL, R. E., DUFFEY, G. H. and PARLIN, R. B., Chem. Rev. **45**, 69 (1949).
3 DABORA, E. K., NICHOLLS, J. A. and MORRISON, R. B., 10th Symposium on Combustion, 817, 1965, Combustion Institute.
4 WOOD, W. W. and KIRKWOOD, J. G., J. Chem. Phys. **22**, 1920 (1954).
5 SOMMERS, W. P., ARS J. **31**, 1780 (1961).
6 SOMMERS, W. P. and MORRISON, R. B., Phys. Fluids **5**, 241 (1962).
7 SICHEL, M., AIAA J. **4**, 264 (1966).
8 HAYES, W. D. and PROBSTEIN, R. F., Hypersonic Flow Theory, Chap III, 1959, Academic Press.
9 NAHMANI, G. and MANHEIMER, Y., J. Chem. Phys. **24**, 1074 (1956).
10 CHAMPBELL, S. W., MALIN, M. E. and HOLLAND, T. E., Detonation in homogeneous explosives, Second ONR Symposium on Detonation, 336, 1955.
11 TSUGÉ, S. FURUKAWA, H., MATSUKAWA, M. and NAKAGAWA, T., Astronautica Acta **15**, 377 (1970).
12 FUJIWARA, T. and TSUGÉ, S., J. Phys. Soc. Japan, **33**, 237 (1972).
13 FUJIWARA, T., Nensho Kenkyu **22** (in Japanese), 8 (1969).
14 KIRKWOOD, J. G. and WOOD, W. W., J. Chem. Phys. **22**, 1915 (1954).
15 HIRSCHFELDER, J. O. and CURTISS, C. F., J. Chem. Phys. **28**, 1130 (1958).
16 BENSON, W. W. and AXWORTHY, A. E., J. Chem. Phys. **26**, 1718 (1957).
17 VOITSEKOVSKII, B. V., MITROFANOV, V. V. and TOPCHIYAN, M. E., 12th Symposium on Combustion, 829, 1969, Combustion Institute.
18 MUNDAY, G., UBBELOHDE, A. R. and WOOD, I. W., Proc. Roy. Soc. **A306**, 159, 171, 179; **A303**, 397 (1968).
19 PARFANOV, A. K., and APIN, A. YA., Combustion, Explosion and Shock Waves **1**, 81 (1965).
20 BELLES, F. E. and EHLERS, J. G., ARS J. **32**, 215 (1962).
21 HARPER, S. A. and GORDON, W. E., Ozone Chemistry and Technology, Advances in Chemistry, 1959, American Chemical Society.
22 ZELDOVICH, IA. B. and KOMPANEETS, A. S., Theory of Detonation, 145, 1960, Academic Press.

Eingereicht am 29. 1. 1973

Anschriften: SHUN-I-CHI TSUGÉ, NASA Ames Research Center, Moffett Field, Calif. 94035, U.S.A.
TOSHITAKA FUJIWARA, Department of Aeronautics, Nagoya University, Nagoya 464, Japan

ZAMM 54, 165 – 171 (1974)

K. GERSTEN

Die Verdrängungsdicke bei Grenzschichten höherer Ordnung

Es werden die von Lighthill stammenden vier Definitionen der Verdrängungsdicke auf kompressible ebene und axialsymmetrische Grenzschichten höherer Ordnung erweitert. Drei Definitionen sind gleichwertig und liefern nur für inkompressible Strömung an undurchlässigen Körpern gleiche Ergebnisse wie die vierte Definition. Für die asymptotische Entwicklung der Verdrängungsdicke bei großen Reynoldszahlen werden die ersten beiden Glieder, also insbesondere die Verdrängungsdicke zweiter Ordnung, berechnet.

The four different definitions of the displacement thickness after Lighthill have been extended to compressible two-dimensional and axisymmetric higher order boundary layers. Three definitions are equivalent and yield the same result as the fourth one only in incompressible flow past impermeable bodies. The first two terms of the asymptotic expansion of the displacement thickness for high Reynolds numbers have been determined, especially the second order displacement thickness.

Производится обобщение четырёх определений толщины вытеснения Ляйтхилла на сжимаемые плоские и осесимметрические пограничные слои высшего порядка. Три определения равноценны. Для несжимаемого течения около непроницаемых тел они приводят к таким-же результатам как и четвёртое определение. Для асимптотического разложения толщины вытеснения при больших числах Рейнольдса определяются первые два члена т. е. толщина вытеснения второго порядка.

A. Einleitung

Von LIGHTHILL [1] wurden für inkompressible PRANDTLsche Grenzschichten vier verschiedene Definitionen der Verdrängungsdicke angegeben, die in ebener Strömung auf dieselben Werte führen.

Im folgenden sollen diese Definitionen auf Grenzschichten höherer Ordnung erweitert werden, und zwar für kompressible ebene und axialsymmetrische Strömungen. Grundlage bildet die Methode der angepaßten asymptotischen Entwicklungen (matched asymptotic expansions) mit einer äußeren Entwicklung (Außenströmung) und einer inneren Entwicklung (Grenzschichtströmung), deren erste Glieder die PRANDTLsche Grenzschichtlösung und deren zweite Glieder die Grenzschichteffekte zweiter Ordnung infolge Krümmung, Verdrängung und Entropiegradienten (infolge gekrümmter Verdichtungsstöße) liefern. Die letztgenannten Effekte zweiter Ordnung auf die Verdrängungsdicke, d. h. die Verdrängungsdicke zweiter Ordnung, sollen hier ermittelt werden. Dabei wird homogene Anströmung vorausgesetzt, und die bei stark verdünnten Gasen auftretenden Sprünge von Geschwindigkeit und Temperatur an der Wand seien vernachlässigt.

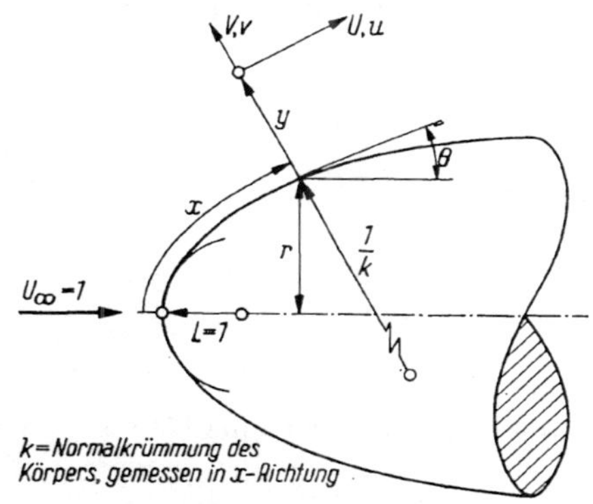

Bild 1. Koordinatensystem
k = bezogene örtliche Krümmung der Meridianlinie

Es wird das Koordinatensystem nach Bild 1 verwendet. Alle Längen sind auf den Krümmungsradius im Staupunkt oder eine andere geeignete Länge, alle übrigen Größen auf die entsprechenden Größen der Anströmung bezogen.

B. Verallgemeinerung der Definitionen von Lighthill

1. Aequivalente Quellen (Equivalent Sources)

Es werden zwei Strömungen um den gegebenen Körper betrachtet, eine *äußere* Strömung mit den Geschwindigkeitskomponenten $U(x, y)$, $V(x, y)$ und der Dichte $R(x, y)$ und eine *innere* Strömung (Grenzschicht) mit $u(x, y)$, $v(x, y)$ und $\varrho(x, y)$. Für beide gilt die Kontinuitätsgleichung

$$\frac{\partial}{\partial x}[(r + y \cos\theta)^j R\, U] + \frac{\partial}{\partial y}[(1 + k\, y)(r + y \cos\theta)^j R\, V] = 0\,, \tag{1}$$

$$\frac{\partial}{\partial x}[(r + y \cos\theta)^j \varrho\, u] + \frac{\partial}{\partial y}[(1 + k\, y)(r + y \cos\theta)^j \varrho\, v] = 0\,, \tag{2}$$

mit $j = 0$ für ebene und $j = 1$ für axialsymmetrische Strömung.

Unter der Annahme, daß die beiden Strömungen in großem Wandabstand ($y \to \infty$) übereinstimmen, liefert Differenzbildung der Gleichungen (1) und (2) und Integration über y von 0 bis ∞

$$r^j R(x, 0)\, V(x, 0) = \frac{dI}{dx} + r^j \varrho(x, 0)\, v(x, 0)\,, \tag{3}$$

mit

$$I(x) = \int_0^\infty [R(x, y)\, U(x, y) - \varrho(x, y)\, u(x, y)]\, (r + y \cos \theta)^j\, dy\,. \tag{4}$$

Danach entspricht der Differenz der Massenströme zuzüglich der Injektion von Masse in der Grenzschicht einer Verteilung von Quellen auf der Kontur mit der Intensität $R(x, 0)$, $V(x, 0)$. Infolge dieser Quellen ist die Körperkontur keine Stromlinie für die äußere Strömung. Es existiert jedoch in der äußeren Strömung eine *Trennstromlinie*, welche die aus dem Unendlichen kommende Strömung von der Quellströmung trennt. Der Wandabstand dieser Trennstromlinie sei als Verdrängungsdicke $D^*(x)$ definiert.

An jeder Stelle x muß zwischen der Wand und der Trennstromlinie diejenige Masse durchströmen, die bis zu dieser Stelle von der Quellverteilung ausgeblasen wurde. Daraus folgt die Definitionsgleichung

$$\int_0^{D^*} R(x, y)\, U(x, y)\, (r + y \cos \theta)^j\, dy = M(x) = \int_0^x r^j\, R(x, 0)\, V(x, 0)\, dx\,. \tag{5}$$

Dabei zählt x vom Staupunkt oder von einer Vorderkante.

Da D^* eine kleine Größe sein soll, können die Funktionen in Gleichung (5) für kleine Wandabstände entwickelt werden:

$$\begin{aligned} R(x, y) &= R(x, 0) + R_y(x, 0)\, y + \cdots, \\ U(x, y) &= U(x, 0) + U_y(x, 0)\, y + \cdots, \\ (r + y \cos \theta)^j &= r^j \left[1 + j \frac{\cos \theta}{r} y + \cdots \right], \end{aligned} \tag{6}$$

wobei die Indizes partielle Ableitungen bedeuten. Damit läßt sich in Gleichung (5) die Integration über y ausführen, und man erhält die folgende Bestimmungsgleichung für $D^*(x)$:

$$\begin{aligned} r^j\, R(x, 0)\, U(x, 0)\, D^*(x) + \frac{r^j}{2} \left[j\, R(x, 0)\, U(x, 0) \frac{\cos \theta}{r} + R(x, 0)\, U_y(x, 0) + U(x, 0)\, R_y(x, 0) \right] \times \\ \times D^{*2}(x) + O(D^{*3}) = M(x) = I(x) + \int_0^x r^j\, \varrho(x, 0)\, v(x, 0)\, dx\,. \end{aligned} \tag{7}$$

Diese allgemeine Beziehung erfordert für den Staupunkt $U = u = 0$ eine Sonderbetrachtung. In Staupunktnähe gilt

$$\begin{aligned} k(x) &= 1 + O(x^2)\,, & r(x) &= x + O(x^3)\,, \\ \cos \theta(x) &= x + O(x^3)\,, & U(x, y) &= U_1(y)\, x + O(x^3)\,, \\ R(x, y) &= R(0, y) + O(x^2)\,, & V(x, 0) &= V(0, 0) + O(x^2)\,. \end{aligned} \tag{8}$$

Damit folgt aus Gleichung (7) der von x unabhängige Wert $D_0^* = D^*(0)$

$$\begin{aligned} R(0, 0)\, U_1(0)\, D_0^* + \frac{1}{2} \big(j\, R(0, 0)\, U_1(0) + R(0, 0)\, U_{1y}(0) + U_1(0)\, R_y(0, 0)\big)\, D_0^{*2} + O(D_0^{*3}) = \\ = \frac{1}{j + 1} R(0, 0)\, V(0, 0)\,. \end{aligned} \tag{9}$$

2. Ersatzströmung (Flow Reduction)

Nach Bild 2 wird die innere Strömung (Bild 2a) durch den Teil der äußeren Strömung oberhalb $D^*(x)$ (Bild 2b) ersetzt, und zwar so, daß diese Ersatzströmung den Massenstrom der inneren Strömung abzüglich des Ausblasemassenstroms besitzt. Es gilt also

$$\int_{D^*}^\infty R(x, y)\, U(x, y)\, (r + y \cos \theta)^j\, dy = \int_0^\infty \varrho(x, y)\, u(x, y)\, (r + y \cos \theta)^j\, dy - \int_0^x r^j\, \varrho(x, 0)\, v(x, 0)\, dx\,. \tag{10}$$

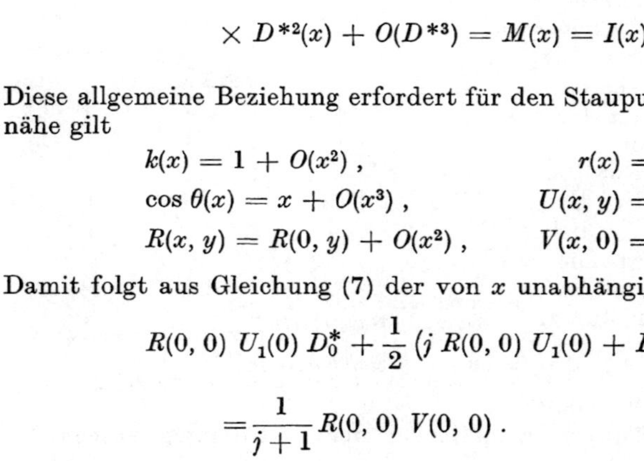

Bild 2. Zur Definition der Verdrängungsdicke $D^*(x)$
a) Innere Strömung
b) Äußere Ersatzströmung

Durch Umformen folgt

$$\int_0^{D^*} R(x, y)\, U(x, y)\, (r + y \cos \theta)^j\, dy = I(x) + \int_0^x r^j\, \varrho(x, 0)\, v(x, 0)\, dx\,, \tag{11}$$

was nach Entwicklung für kleine D^* wieder Gleichung (7) ergibt. Bei undurchlässigen ebenen Körpern müssen die beiden schraffierten Flächen in Bild 2b gleich sein.

3. Massenstromvergleich (Velocity Comparison)

Da $D^*(x)$ die Lage der Trennstromlinie beschreibt, kann sofort die kinematische Strömungsbedingung angesetzt werden:

$$\frac{dD^*}{dx} = \frac{(1 + k D^*) V(x, D^*)}{U(x, D^*)} . \tag{12}$$

Wieder kann man die Funktionen $V(x, D^*)$ und $U(x, D^*)$ für kleine Wandabstände entsprechend Gleichung (6) entwickeln. Die dabei auftretenden partiellen Ableitungen $V_y(x, 0)$ und $V_{yy}(x, 0)$ lassen sich mit Hilfe von (1) bzw. mit der aus (1) durch Ableitung nach y entstehenden Gleichung wie folgt durch andere Größen ersetzen:

$$r^j R V_y = - \bar{R}_y V - \frac{d}{dx} (r^j R U) , \tag{13}$$

$$r^j R V_{yy} = - \bar{R}_{yy} V - 2 \bar{R}_y V_y - \frac{d}{dx} (j \cos \theta R U + r^j (R_y U + R U_y)) , \tag{14}$$

wobei alle Funktionen bei $y = 0$ zu nehmen sind. Zur Abkürzung wurde dabei gesetzt

$$\bar{R}(x, y) = (1 + k y) (r + y \cos \theta)^j R(x, y) . \tag{15}$$

Damit ergibt sich aus Gleichung (12) schließlich

$$\frac{d}{dx} (r^j R U D^*) - r^j R V = H(x) = (r^j k R - \bar{R}_y) V D^* - r^j R U_y D^* \frac{dD^*}{dx} + \\ + r^j R \left(k V_y + \frac{1}{2} V_{yy}\right) D^{*2} + O(D^{*3}) , \tag{16}$$

wobei wieder alle Funktionen bei $y = 0$ genommen worden sind. Integration über x liefert dann

$$r^j R U D^* = M(x) + \int_0^x H(x) \, dx + O(D^{*3}) . \tag{17}$$

Dieses Ergebnis ist mit Gleichung (7) identisch, wie sich durch eine kurze Rechnung zeigen läßt. Damit gilt auch (9) für D_0^*. Man kann jedoch auch direkt eine Beziehung für D_0^* erhalten. Im Staupunkt ergibt sich in (12) ein unbestimmter Ausdruck. Es muß also gelten

$$V(0, D_0^*) = 0 , \tag{18}$$

oder nach Entwicklung

$$V(0, 0) + V_y(0, 0) D_0^* + \frac{1}{2} V_{yy}(0, 0) D_0^{*2} + O(D_0^{*3}) = 0 , \tag{19}$$

was wegen der Gleichungen (13) und (14) mit Gleichung (9) identisch ist.

4. Schwerpunkt der Drehung (Mean Vorticity)

Die Drehung der *äußeren* und *inneren* Strömung ist wie folgt definiert:

$$2 \Omega(x, y) = \frac{1}{1 + k y} \left\{ \frac{\partial V}{\partial x} - \frac{\partial}{\partial y} ((1 + k y) U) \right\} , \tag{20}$$

$$2 \omega(x, y) = \frac{1}{1 + k y} \left\{ \frac{\partial v}{\partial x} - \frac{\partial}{\partial y} ((1 + k y) u) \right\} . \tag{21}$$

Im gewählten Koordinatensystem sind also die Formeln für den ebenen und axialsymmetrischen Fall gleich.

Die Verdrängungsdicke kann jetzt als Wandabstand des Schwerpunktes der Verteilung $\Omega(x, y) - \omega(x, y)$ definiert werden. Es gilt also

$$D^*(x) = \frac{\int_0^\infty (\Omega - \omega) \, y \, (r + y \cos \theta)^j \, dy}{\int_0^\infty (\Omega - \omega) (r + y \cos \theta)^j \, dy} . \tag{22}$$

Einsetzen liefert mit der zusätzlichen Annahme, daß die innere Lösung die Haftbedingung

$$u(x, 0) = 0 \tag{23}$$

erfüllen soll,

$$D^*(x) = \frac{\int\limits_0^\infty (U-u)\,(r+y\cos\theta)^j\,dy + \int\limits_0^\infty \left(\frac{\partial}{\partial x}(V-v) - k\,(U-u)\right)\frac{y\,(r+y\cos\theta)^j\,dy}{1+k\,y}}{r^j\,U(x,0) + \int\limits_0^\infty \left(\frac{\partial}{\partial x}(V-v) - k\,(U-u)\right)(r+y\cos\theta)^j\,\frac{dy}{1+k\,y}}\,. \tag{24}$$

Da die Dichte hierbei nicht auftritt, kann diese Definition nur für inkompressible Strömung auf die gleichen Werte führen wie die übrigen drei Definitionen.

Es sei angemerkt, daß in allen hier entwickelten Bestimmungsgleichungen die Begriffe *Grenzschichtdicke* und *Außenrand der Grenzschicht* nicht vorkommen, wodurch sich eine in der Literatur [6], [7], [8] mehrfach diskutierte Streitfrage erübrigt.

C. Asymptotische Entwicklungen

Die beiden Lösungen seien von der Reynoldszahl Re abhängig und mögen asymptotische Entwicklungen für den Störparameter

$$\varepsilon = \frac{1}{\sqrt{Re}} = \frac{1}{\sqrt{\dfrac{L\,U_\infty}{\nu_\infty}}} \tag{25}$$

besitzen. Dann kann auch die Verdrängungsdicke $D^*(x)$ asymptotisch entwickelt werden. Es gelte:

$$\begin{aligned}
D^*(x) &= \varepsilon\,D_1^*(x) + \varepsilon^2\,D_2^*(x) + O(\varepsilon^3)\,,\\
D_0^* &= \varepsilon\,D_{01}^* + \varepsilon^2\,D_{02}^* + O(\varepsilon^3)\,,\\
I(x) &= \varepsilon\,I_1(x) + \varepsilon^2\,I_2(x) + O(\varepsilon^3)\,,\\
U(x,y) &= U_1(x,y) + \varepsilon\,U_2(x,y) + O(\varepsilon^2)\,,\\
V(x,y) &= V_1(x,y) + \varepsilon\,V_2(x,y) + O(\varepsilon^2)\,,\\
R(x,y) &= R_1(x,y) + \varepsilon\,R_2(x,y) + O(\varepsilon^2)\,,\\
\bar{R}(x,y) &= \bar{R}_1(x,y) + \varepsilon\,\bar{R}_2(x,y) + O(\varepsilon^2)\,,\\
u(x,y) &= u_1(x,N) + \varepsilon\,u_2(x,N) + O(\varepsilon^2)\,,\\
v(x,y) &= \varepsilon\,v_1(x,N) + \varepsilon^2\,v_2(x,N) + O(\varepsilon^3)\,,\\
\varrho(x,y) &= \varrho_1(x,N) + \varepsilon\,\varrho_2(x,N) + O(\varepsilon^2)\,,
\end{aligned} \tag{26}$$

mit der inneren Variablen

$$N = \frac{y}{\varepsilon}. \tag{27}$$

Aus Gleichung (4) ergibt sich dann durch Einsetzen, Entwickeln der äußeren Lösung für kleine Werte y und Ordnen nach Potenzen von ε:

$$I_1(x) = r^j \int\limits_0^\infty \{R_1(x,0)\,U_1(x,0) - \varrho_1(x,N)\,u_1(x,N)\}\,dN\,, \tag{28a}$$

$$\begin{aligned}
I_2(x) = r^j \int\limits_0^\infty \Big\{&R_2(x,0)\,U_1(x,0) + R_1(x,0)\,U_2(x,0) + \{R_{1y}(x,0)\,U_1(x,0) + R_1(x,0)\,U_{1y}(x,0)\}\,N -\\
&- \varrho_2(x,N)\,u_1(x,N) - \varrho_1(x,N)\,u_2(x,N) + j\,\frac{\cos\theta}{r}\,\big(R_1(x,0)\,U_1(x,0) - \varrho_1(x,N)\,u_1(x,N)\big)N\Big\}\,dN\,.
\end{aligned} \tag{28b}$$

Aus den Gleichungen (7) und (17) erhält man das gleiche Ergebnis für $D_1^*(x)$:

$$r^j\,R_1(x,0)\,U_1(x,0)\,D_1^*(x) = I_1(x) + \int\limits_0^x r^j\,\varrho_1(x,0)\,v_1(x,0)\,dx\,. \tag{29}$$

Dabei wird in (17) noch benutzt, daß nach [2] gilt

$$V_1(x,0) = 0\,. \tag{30}$$

Das Ergebnis in (29) wurde bereits von Hayasi [3] angegeben.

Im Staupunkt gilt mit

$$U_1(x,0) = U_{11}\,x + O(x^2)\,, \tag{31}$$

$$R_1(0,0)\,U_{11}\,D_{01}^* = \lim_{x\to 0}\frac{I_1(x)}{x^{j+1}} + \frac{1}{j+1}\,\varrho_1(0,0)\,v_1(0,0)\,. \tag{32}$$

Aus (24) erhält man (29), wenn $R_1(x, 0) = \varrho_1(x, N) = 1$ und $v_1(x, 0) = 0$ gesetzt wird.
Für die Verdrängungsdicke zweiter Ordnung erhält man nach den ersten drei Definitionen

$$r^j R_1(x, 0)\, U_1(x, 0)\, D_2^*(x) = I_2(x) + \int_0^x r^j \big(\varrho_1(x, 0)\, v_2(x, 0) + \varrho_2(x, 0)\, v_1(x, 0)\big)\, dx - \\ - r^j \big(R_1(x, 0)\, U_2(x, 0) + R_2(x, 0)\, U_1(x, 0)\big)\, D_1^* + A(x)\,. \tag{33}$$

Für das Zusatzglied erhält man aus den Gleichungen (7) bzw. (17) die folgenden gleichwertigen Ausdrücke:

$$A(x) = -\frac{r^j}{2}\left(j\, R_1(x, 0)\, U_1(x, 0)\frac{\cos\theta}{r} + R_1(x, 0)\, U_{1y}(x, 0) + U_1(x, 0)\, R_{1y}(x, 0)\right) D_1^{*2} \\ = -\int_0^x r^j \left\{R_1(x, 0)\, V_2(x, 0)\, D_1^*(x) - k\, U_1(x, 0) - U_{1y}(x, 0)\, R_1(x, 0)\, D_1^*(x)\frac{dD_1^*}{dx}\right\} dx\,. \tag{34}$$

Aus Gleichung (24) erhält man ein Ergebnis gleicher Form, wenn inkompressible Strömung ohne Ausblasen oder Absaugen vorausgesetzt wird. Für das Zusatzglied $A(x)$ ergibt sich dabei jedoch der Ausdruck

$$A(x) = k\, r^j\, U_1(x, 0)\, D_1^{*2} - k \int_0^\infty r^j \big(U_1(x, 0) - u_1(x, N)\big)\, N\, dN\,, \tag{35}$$

der im allgemeinen nicht mit dem auf inkompressible Strömung ohne Ausblasen oder Absaugen spezialisierten $A(x)$ nach Gleichung (34) übereinstimmt. Das ist auch nicht zu erwarten, da bekanntlich die Außenströmung dritter Ordnung keine Potentialströmung ist und sich daher auch nicht mehr durch eine Wirbelbelegung auf der Trennungsstromlinie darstellen läßt. Für den Staupunkt erhält man mit

$$U_2(x, 0) = U_{21}\, x + O(x^2)\,, \\ U_1(x, y) = [U_{11} + U_{11y}\, y + O(y^2)]\, x + O(x^2)\,, \tag{36}$$

die Verdrängungsdicke zweiter Ordnung

$$R_1(0, 0)\, U_{11}\, D_{02}^* = \lim_{x\to 0}\frac{I_2(x)}{x^{j+1}} + \frac{1}{j+1}\big(\varrho_1(0, 0)\, v_2(0, 0) + \varrho_2(0, 0)\, v_1(0, 0)\big) - \\ - \big(R_1(0, 0)\, U_{21} + R_2(0, 0)\, U_{11}\big)\, D_{01}^* + \lim_{x\to 0}\frac{A(x)}{x^{j+1}}, \tag{37}$$

wobei gilt

$$\lim_{x\to 0}\frac{A(x)}{x^{j+1}} = -\frac{1}{2}\big(j\, R_1(0, 0)\, U_{11} + R_1(0, 0)\, U_{11y} + U_{11}\, R_{1y}(0, 0)\big)\, D_{01}^{*2}\,. \tag{38}$$

Aus Gleichung (35) erhält man für den Staupunkt

$$\lim_{x\to 0}\frac{A(x)}{x^{j+1}} = U_{11}\, D_{01}^{*2} - U_{11}\lim_{x\to 0}\int_0^\infty \left(1 - \frac{u_1(x, N)}{U_1(x, 0)}\right) N\, dN\,. \tag{39}$$

D. Wirbelsatz von Crocco

Nach [2] erfüllen die beiden ersten Glieder der asymptotischen Entwicklung der äußeren Strömung die Gleichungen der reibungslosen Strömung. Daher gilt für sie der Wirbelsatz von CROCCO. Für $y = 0$ folgt daraus nach [2]

$$U_{1y}(x, 0) = -k\, U_1(x, 0) - r^j R_1(x, 0)\, T_1(x, 0)\, S_1'(0)\,. \tag{40}$$

Dabei gilt für den Entropiegradienten an der Wand

$$S_1'(0) = \left(\frac{dS_1}{d\psi_1}\right)_{y=0} = \frac{-4(\gamma-1)\,(Ma_\infty^2 - 1)^2\, a^{1+j}}{2\gamma\, Ma_\infty^2 - (\gamma-1)\,\big(2 + (\gamma-1)\, Ma^2\big)}\, j\,, \tag{41}$$

wobei γ der Isentropenexponent und a das Verhältnis von Körper zu Stoßradius im Staupunkt sind. Dieser Entropiegradient kann nur bei einem gekrümmten Stoß vor dem Körper, also nur für $Ma_\infty > 1$ auftreten. Nach Gleichung (41) ist $S_1'(0)$ nur im axialsymmetrischen von Null verschieden, während es für $j = 0$ verschwindet. Danach sind bei allen hier betrachteten ebenen und axialsysmmetrischen Strömungen die äußeren Lösungen erster Ordnung auf der Kontur drehungsfrei. Einzige Ausnahme bildet die axialsymmetrische Überschallströmung um stumpfe Körper. Mit Hilfe von Gleichung (40) kann das Glied $U_{1y}(x, 0)$ in Gleichung (34) ersetzt werden.

E. Beispiele

1. Ebene inkompressible Staupunktströmung

Nach [4] wird für die innere Lösung der Ansatz

$$u(x, \eta) = U_{11}\, x \left(f'(\eta) + \frac{\varepsilon}{\sqrt{U_{11}}} F'_c(\eta) + \varepsilon \frac{U_{21}}{U_{11}} F'_d(\eta) + O(\varepsilon^2) \right) \tag{42}$$

gewählt, wobei

$$\eta = \sqrt{U_{11}}\, N \tag{43}$$

die übliche Grenzschichtvariable ist.

Für die Normalgeschwindigkeit an der Wand gilt dann

$$\begin{aligned} v_1(0, 0) &= -\sqrt{U_{11}}\, f(0)\,, \\ v_2(0, 0) &= -\left[F_c(0) + \frac{U_{21}}{\sqrt{U_{11}}} F_d(0) \right]. \end{aligned} \tag{44}$$

Damit ergeben sich die Verdrängungsdicken zu

$$D^*_{01} = \frac{1}{\sqrt{U_{11}}} \lim_{\eta \to \infty} (\eta - f(\eta)) \tag{45}$$

und

$$D^*_{02} = \frac{-1}{U_{11}} \lim_{\eta \to \infty} \left(\frac{\eta^2}{2} + F_c(\eta) \right) + \frac{U_{21}}{U_{11}^{3/2}} \lim_{\eta \to \infty} (\eta - F_d(\eta)) - \frac{U_{21}}{U_{11}} D^*_{01} + \frac{1}{2} D^{*2}_{01}\,. \tag{46}$$

In [4] sind Zahlenwerte der Grenzwerte für verschiedene Anfangswerte $f(0)$ bei $F_c(0) = F_d(0) = 0$ angegeben. Danach ist es durch Absaugen stets möglich, die Verdrängungsdicken zum Verschwinden zu bringen.

Es verschwindet D^*_{01} für $f(0) = 0{,}536$. Durch eine geeignete Absaugeverteilung längs der Körperkontur kann die gesamte Verdrängungsströmung vermieden werden, so daß $U_{21} = 0$ gilt. Dann würde für $F_c(0) = 0{,}43$ auch D^*_{02} verschwinden.

2. Achsenparallele kompressible Strömung an einem Kreiszylinder ohne Verdrängungseffekt

Diese Strömung ist in [5] für ideales Gas konstanter spezifischer Wärme mit der Viskosität proportional der Temperatur und mit der Prandtl-Zahl $Pr = 0{,}72$ behandelt worden. Danach hat die innere Lösung für $D^*_1(x) = 0$ die Geschwindigkeitsverteilung

$$u(x, \eta) = \frac{1}{2} f'(\eta) + \frac{\varepsilon}{2} \sqrt{x}\, F'(\eta) + O(\varepsilon^2) \tag{47}$$

und die Temperaturverteilung

$$T(x, \eta) = \varrho^{-1}(x, \eta) = \frac{1}{2} g'(\eta) + \frac{\varepsilon}{2} \sqrt{x}\, G'(\eta) + O(\varepsilon^2) \tag{48}$$

mit der Grenzschichtvariablen

$$\eta = \frac{1}{2\sqrt{x}} \int_0^N \varrho\, (1 + \varepsilon N)\, dN\,. \tag{49}$$

Für den Massenstrom an der Wand gilt dann

$$\varrho_1(x, 0)\, v_1(x, 0) = -\frac{1}{2\sqrt{x}} f(0)\,, \tag{50}$$

$$\varrho_1(x, 0)\, v_2(x, 0) = -F(0)\,. \tag{51}$$

Damit erhält man formal die Verdrängungsdicke erster Ordnung zu

$$D^*_1(x) = \sqrt{x} \lim_{\eta \to \infty} \{g(\eta) - f(\eta)\}\,. \tag{52}$$

Nach [5] verschwindet $D^*_1(x)$ im Bereich $|f(0)| < 0{,}1$ für

$$T_W = \frac{0{,}110 + 0{,}087\, f(0)}{0{,}970 - 0{,}940\, f(0)} - \frac{0{,}058 - 0{,}033\, f(0)}{0{,}970 - 0{,}940\, f(0)} Ma^2_\infty\,. \tag{53}$$

Also auch bei undurchlässiger Wand, d. h. für $f(0) = 0$ kann bei genügend geringen Wandtemperaturen die Verdrängungsdicke erster Ordnung verschwinden.

Die Verdrängungsdicke zweiter Ordnung vereinfacht sich wegen $D_1^*(0) = 0$ zu

$$D_2^*(x) = x \lim_{\eta \to \infty} \left(G(\eta) - F(\eta)\right) . \tag{54}$$

Durch geeignete Wahl von $F(0)$ nach Gl. (51), also durch Absaugen mit konstanter, von x unabhängiger Geschwindigkeit, verschwindet auch $D_2^*(x)$.

Literatur

1 LIGHTHILL, M. J., On displacement thickness. J. Fluid Mech. **4**, 383–392 (1958).
2 VAN DYKE, M., Second-order compressible boundary layer theory with application to blunt bodies in hypersonic flow. Progress in Astronautics and Rocketry Vol. 7: Hypersonic Flow Research (Editor: F. R. Riddell), 1962, 37–76.
3 HAYASI, N., Displacement thickness of the boundary layer with blowing. AIAA Journal. **3**, 2348–2349 (1965).
4 GERSTEN, K., GROSS, J. F., BÖRGER, G.-G., Die Grenzschicht höherer Ordnung an der Staulinie eines schiebenden Zylinders mit starkem Absaugen oder Ausblasen. Z. Flugwiss. **20**, 330–341 (1972).
5 GERSTEN, K., GROSS, J. F., The second-order boundary layer along a circular cylinder in supersonic flow. Erscheint in Int. J. Heat Mass Transfer.
6 FANNELÖP, T. K., Displacement thickness for boundary layers with surface mass transfer. AIAA Journal **4**, 1142–1145 (1966).
7 BURGGRAF, O. R., Choice of boundary conditions in viscous interaction theory. AIAA Journal **4**, 1145–1147 (1966).
8 LEWIS, C., Comments on "Displacement thickness for boundary layers with surface mass transfer". AIAA Journal **5**, 831–832 (1967).

Eingereicht am 8. 2. 1973

Anschrift: Prof. Dr.-Ing. K. GERSTEN, Institut für Thermo- und Fluiddynamik der Ruhr-Universität Bochum, 4630 Bochum, Postfach 2148, BRD

ZAMM 54, 173 – 179 (1974)

P. Y. WILLEMS / Y. NISHIKAWA

Kinematical Drift of two Degree of Freedom Gyroscopes

Es wird das Verhalten von Kreiseln mit zwei Freiheitsgraden untersucht, die einer monofrequenten Schwingung unterworfen sind. Mit Hilfe einer Methode, die sich für die Untersuchung der Stabilität schwach nichtlinearer dynamischer Systeme eignet, werden nichtlineare und parametrische Resonanzen analysiert. Es werden Stabilitätsbedingungen angegeben, die die physikalischen Parameter des Systems enthalten.

The behaviour of two degrees of freedom gyroscopes subject to monofrequency vibration is investigated. Nonlinear and parametrical resonances are analyzed by using a method suitable for stability investigation of slightly nonlinear dynamical systems. Stability conditions are provided in terms of the physical parameters of the system.

Исследуется поведение гироскопов, подвержённых воздействию моночастотного колебания. С помощью метода, пригодного для исследования стабильности слабо нелинейных динамических систем, производится анализ нелинейных и параметрных резонансов. Приводятся условия стабильности, содержащие физические параметры системы.

The kinematical drift of a rigid two degrees of freedom gyroscope subject to monofrequency vibrations will be investigated as an application of the method for stability investigation of slightly nonlinear dynamic systems with several degrees of freedom.

This problem is well-known and is formulated in the literature. A first attend to point out the influence of vibrations on the behaviour of a rotating body was made by DE JONQUIÈRES as early as in 1886 [1].

In the mid-1940's troubles caused by vibration were observed in inertial instruments. According to CRAWLEY and MAUNDER [2] an unpublished report was written on the subject by WEIDMANN and SEEBACH in 1944. This report "contains a preliminary but rigourous analysis of the vibrational drift of a free balanced gyroscope, together with some experimental results". Two theoretical papers on similar problems appeared in 1955, [3], [4].

The actual problem is clearly stated by MAUNDER [5] and a solution is given by CRAWLEY and MAUNDER [2]. The problem of drift compensation is also investigated in [6]. The motion and the drift of a gyroscope with CARDAN's suspension and nonaxially placed rotor was investigated by KLIMOV for a system with fiction [7] and a system without any damping mechanism [8]. In fact, methods quite similar to those presented here are used in these papers.

On the other hand the gyroscope drift associated with vehicle vibration is investigated by WENGLANZ and KANE [9].

Some inertial reference frame $\{\hat{\boldsymbol{x}}_1, \hat{\boldsymbol{x}}_2, \hat{\boldsymbol{x}}_3\}$ coincides with the gimbals axes of the gyroscope in its canonical position (gimbals perpendicular to each other). Without lack of generality, it will be assumed that the vibration occurs about the $\hat{\boldsymbol{x}}_3$-axis (Fig. 1). The support frame will be the $\{\hat{\boldsymbol{a}}_1, \hat{\boldsymbol{a}}_2, \hat{\boldsymbol{a}}_3\}$-frame, and the $\hat{\boldsymbol{a}}_3$-axis will coincide with $\hat{\boldsymbol{x}}_3$. If the $\hat{\boldsymbol{a}}_1$-axis is aligned with the outer gimba laxis, an angular rotation θ_1 about $\hat{\boldsymbol{a}}_1$ will bring the support frame in coincidence with the outer gimbal frame, $\{\hat{\boldsymbol{b}}_1, \hat{\boldsymbol{b}}_2, \hat{\boldsymbol{b}}_3\}$. Similarly, if the $\hat{\boldsymbol{b}}_2$-axis of this frame is aligned with the inner gimbal axis, a rotation θ_2 about the $\boldsymbol{b}_2$-axis will bring the outer gimbal frame in coincidence with the inner gimbal frame $\{\hat{\boldsymbol{c}}_1, \hat{\boldsymbol{c}}_2, \hat{\boldsymbol{c}}_3\}$, the $\hat{\boldsymbol{c}}_3$-axis of this frame being parallel to the wheel axis (Fig. 2). The inner gimbal and the wheel, will be considered as one gyrostat. This supposes a symmetric wheel.

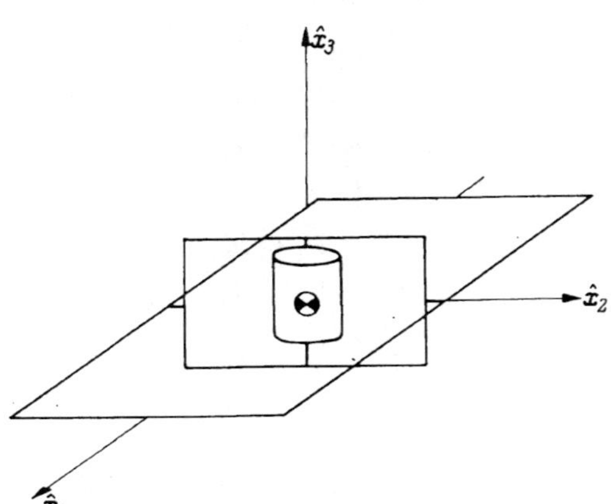

Fig. 1

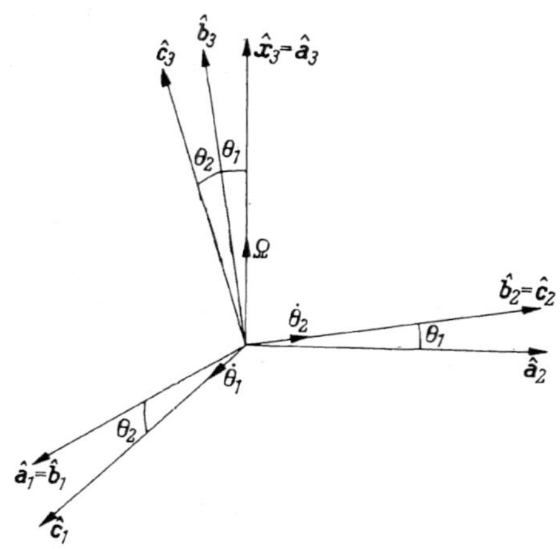

Fig. 2

It will be assumed that the outer gimbal and the gyrostat have the same center of mass, the origin of the above defined reference frame.

The rotational equations for the gyrostat are the well-known LIOUVILLE's equations

$$\boldsymbol{L} = \boldsymbol{I} \cdot \dot{\boldsymbol{\omega}} + \boldsymbol{\omega} \times \boldsymbol{I} \cdot \boldsymbol{\omega} + \boldsymbol{\omega} \times \boldsymbol{H}_0 , \tag{2.1}$$

where

$\boldsymbol{L}$ is the external torque applied around the center of mass by the outer gimbal,
$\boldsymbol{H}_0$ the internal relative angular momentum,
$\boldsymbol{\omega}$ the angular velocity of gyrostat reference frame,
$\boldsymbol{I}$ the inertia dyadic of the gyrostat.

The vibration will be a monofrequency vibration of known amplitude, $\lambda \cos \omega t$. The angular velocity of the vibration being $\Omega = -\lambda \omega \sin \omega t$, the vector $\boldsymbol{\omega}$ is then given by

$$\boldsymbol{\omega} = \Omega \,\hat{\boldsymbol{a}}_3 + \dot{\theta}_1 \,\hat{\boldsymbol{a}}_1 + \dot{\theta}_2 \,\hat{\boldsymbol{b}}_2 . \tag{2.2}$$

The equations (2.1) will be expressed in the $\{\hat{\mathbf{c}}_1, \hat{\mathbf{c}}_2, \hat{\mathbf{c}}_3\}$-frame and

$$\boldsymbol{L} = L_\alpha \,\hat{\mathbf{c}}_\alpha , \qquad \boldsymbol{\omega} = \omega_\alpha \,\hat{\mathbf{c}}_\alpha , \tag{2.3}$$

where from kinematics:

$$\begin{aligned} \omega_1 &= \dot{\theta} \cos \theta_2 - \Omega \cos \theta_1 \sin \theta_2 , \\ \omega_2 &= \dot{\theta}_2 + \Omega \sin \theta_1 , \\ \omega_3 &= \dot{\theta}_1 \sin \theta_2 + \Omega \cos \theta_1 \cos \theta_2 . \end{aligned} \tag{2.4}$$

If further, the axes $\hat{\mathbf{c}}_1$, $\hat{\mathbf{c}}_2$ and $\hat{\mathbf{c}}_3$ are the principal axes of the gyrostat, I_1, I_2 and I_3 being the moments of inertia, the vectorial equation (2.1) may be written:

$$\begin{aligned} L_1 &= I_1 \dot{\omega}_1 - (I_2 - I_3)\, \omega_2 \omega_3 + H_0 \omega_2 , \\ L_2 &= I_2 \dot{\omega}_2 - (I_3 - I_1)\, \omega_3 \omega_1 - H_0 \omega_1 , \\ L_3 &= I_3 \dot{\omega}_3 - (I_1 - I_2)\, \omega_1 \omega_2 . \end{aligned} \tag{2.5}$$

The components of $\boldsymbol{L}$ in the $\{\hat{\boldsymbol{b}}_1 \hat{\boldsymbol{b}}_2 \hat{\boldsymbol{b}}_3\}$-frame being given by

$$\boldsymbol{L} = L'_a \,\hat{\boldsymbol{b}}_a \tag{2.6}$$

where

$$L'_1 = L_1 \cos \theta_2 + L_3 \sin \theta_2 , \qquad L'_2 = L_2 , \qquad L_3 = L_3 \cos \theta_2 - L_1 \sin \theta_2 . \tag{2.7}$$

The condition for equilibrium of the outer gimbal requires

$$M_1 = L'_1 + f_1 \dot{\theta}_1 + A \ddot{\theta}_1 - (B - C)\, \Omega^2 \sin \theta_1 \cos \theta_1 , \qquad M_2 = L'_2 + f_2 \dot{\theta}_2 , \tag{2.8}$$

where

A, B and C are the moments of inertia of the outer gimbal about the $\hat{\boldsymbol{b}}_1$-axis,
M_1 and M_2 are the torques applied by the torquers of the outer and inner gimbals respectively,
f_1 and f_3 are the viscous damping coefficients..

If the behaviour of the system around some position of the gimbals is investigated, the equations of motion may be linearized around this position defined by the angles θ_{10} and θ_{20}, and

$$\theta_1 = \theta_{10} + x_1 , \qquad \theta_2 = \theta_{20} + x_2 , \tag{2.9}$$

where x_1 and x_2 are small quantities of the order of λ, the amplitude of the vibration.

Upto the second order in $x_1\, x_2\, \Omega$ and their time derivatives, the equations of motion are

$$\begin{aligned} &(A + I_1 d^2 + I_3 b^2)\, \ddot{x}_1 + f_1 \dot{x}_1 + [H_0 c\, d\Omega + a\, b\, d\, (I_1 - I_3)\, \dot{\Omega}]\, x_1 + \\ &+ [H_0 d + c\, (d^2 - b^2)\, (I_3 - I_1)\, \Omega - c\, I_2\, \Omega]\, \dot{x}_2 + [c\, (b^2 - d^2)\, (I_1 - I_3)\, \dot{\Omega} - H_0 a\, b\, \Omega]\, x_2 + \\ &+ (I_3 - I_1)\, c\, b\, d\, \dot{\Omega} + H_0 a\, d\, \Omega + [(I_3 - I_2)\, d^2 - (I_2 - I_1)\, b^2]\, a\, c\, \Omega^2 - (B - C)\, \Omega^2 a\, c + \\ &+ 2\, (I_3 - I_1)\, b\, d\, \ddot{x}_1 x_2 + 2\, (I_3 - I_1)\, b\, d\, \dot{x}_1 \dot{x}_2 - H_0 b\, x_2 \dot{x}_2 = M_1 , \\ &I_2 \ddot{x}_2 + f_2 \dot{x}_2 + c\, d\, H_0\, \Omega\, x_2 + [c\, (I_1 - I_3)\, (d^2 - b^2)\, \Omega + I_2 c\, \Omega - H_0 d]\, \dot{x}_1 + [I_2 c\, \dot{\Omega} - a\, b\, H_0\, \Omega]\, x_1 + \\ &+ a\, I_2\, \dot{\Omega} - (I_1 - I_3)\, c^2 b\, d\Omega^2 + H_0 c\, b\, \Omega + (I_1 - I_3)\, b\, d\, \dot{x}_1^2 + H_0 b\, \dot{x}_1 x_2 = M_2 , \end{aligned} \tag{2.10}$$

where for writing convenience

$$\begin{aligned} \sin\, \theta_{10} &= a , \qquad & \sin\, \theta_{20} &= b , \\ \cos\, \theta_{10} &= c , \qquad & \cos\, \theta_{20} &= d , \end{aligned} \tag{2.11}$$

It will also be assumed that the differences of moments of inertia are small compared with the moments of inertia. Then, as it may checked in (2.10), the terms containing these differences only slighly affect the behavior of the system and in first approximation they will be neglected.

For λ identically equal to zero, the eigenfrequencies of the linear system without damping are given by

$$\omega_1^2 = 0\,, \qquad \omega_0^2 = \alpha\,\beta\,, \tag{2.12}$$

where

$$\alpha = \frac{H_0\,d}{I}, \qquad \beta = \frac{H_0\,d}{I_2}, \tag{2.13}$$

with $I = A + I_1\,d^2 + I_3\,b^2$.

For damped systems the normalized parameters ξ_1 and ξ_2 will be defined as

$$\xi_1 = \frac{f_1}{\lambda\,\omega\,I}, \qquad \xi_2 = \frac{f_2}{\lambda\,\omega\,I_2}. \tag{2.14}$$

Then defining the vector X such as

$$X^T = [x_1, x_2, \dot{x}_1, \dot{x}_2]\,,$$

the system (2.10) may be written in matrix form as

$$\dot{X} = A\,X + \lambda\,\omega\cos\omega t\,B\,X + \lambda\,\omega\sin\omega t\,C\,X - \\ - \lambda\,\omega\,Z\,X + \lambda\,\omega\cos\omega t\,F_1 + \lambda\,w\sin\omega t\,F_2 + N\,L(X)\,, \tag{2.15}$$

where

$$A = \begin{bmatrix} 0 & 0 & 1 & 0 \\ 0 & 0 & 0 & 1 \\ 0 & 0 & 0 & -\alpha \\ 0 & 0 & \beta & 0 \end{bmatrix}, \qquad B = \begin{bmatrix} 0 & 0 & 0 & 0 \\ 0 & 0 & 0 & 0 \\ -c\,a & \dfrac{a\,b\,\alpha}{d} & 0 & \dfrac{c\,I_2}{I} \\ \dfrac{a\,b\,\beta}{d} & -c\,\beta & -c & 0 \end{bmatrix},$$

$$C = \begin{bmatrix} 0 & 0 & 0 & 0 \\ 0 & 0 & 0 & 0 \\ 0 & 0 & 0 & 0 \\ c\,\omega & 0 & 0 & 0 \end{bmatrix}, \qquad Z = \begin{bmatrix} 0 & 0 & 0 & 0 \\ 0 & 0 & 0 & 0 \\ 0 & 0 & \xi_1 & 0 \\ 0 & 0 & 0 & \xi_2 \end{bmatrix} \tag{2.16}$$

$$F_1 = \begin{bmatrix} 0 \\ 0 \\ -a\,\alpha \\ -\dfrac{c\,b\,\beta}{d} \end{bmatrix}, \qquad F_2 = \begin{bmatrix} 0 \\ 0 \\ 0 \\ a\,\omega \end{bmatrix}, \qquad N\,L(X) = \frac{b}{d}\begin{bmatrix} 0 \\ 0 \\ a\,x_2\,\dot{x}_2 \\ -\beta\,\dot{x}_1\,x_2 \end{bmatrix}.$$

This system may be transformed in a real canonical system using a linear transformation,

$$X = P\,Y\,, \tag{2.17}$$

which transforms $\dot{X} = A\,X$ into $\dot{Y} = \Lambda\,Y$,
where

$$\Lambda = \begin{bmatrix} 0 & 0 & 0 & 0 \\ 0 & 0 & 0 & \omega_0 \\ 0 & 0 & 0 & 0 \\ 0 & -\omega_0 & 0 & 0 \end{bmatrix}. \tag{2.18}$$

Such a transformation matrix P may be defined in terms of the eigenvalues of the system for $\lambda = 0$, [10], and here

$$P = \begin{bmatrix} 1 & 1 & 0 & 0 \\ 0 & 0 & 1 & -\dfrac{\beta}{\omega_0} \\ 0 & 0 & 0 & \omega_0 \\ 0 & \beta & 0 & 0 \end{bmatrix}. \tag{2.19}$$

Using the transformation in (2.15) the following system is obtained

$$\dot{Y} = \Lambda\,Y + \lambda\,\omega\cos\omega t\,G\,Y + \lambda\,\omega\sin\omega t\,H\,Y - \lambda\,\omega\,D\,Y + \lambda\,\omega\cos\omega t\,V_1 + \\ + \lambda\,\omega\sin\omega t\,V_2 + N\,, \tag{2.20}$$

where

$$\begin{aligned} \Lambda &= P^{-1} A P\,, & D &= P^{-1} Z P\,, & G &= P^{-1} B P\,, & H &= P^{-1} C P\,, \\ V_1 &= P^{-1} F_1\,, & V_2 &= P^{-1} F_2\,, & N &= P^{-1} N\, L(P\,Y)\,, \end{aligned} \tag{2.21}$$

If g_{ij}, h_{ij}, d_{ij}, v_{1i}, v_{2i} and n_i are the non zero elements of matrices defined in (2.21) the system (2.20) is equivalent to

$$\begin{aligned} \dot y_1 &= \lambda\,\omega\, g_{11} \cos \omega t\, y_1 + \lambda\,\omega\, g_{12} \cos \omega t\, y_2 + \lambda\,\omega\, g_{13} \cos \omega t\, y_3 + \lambda\,\omega\, g_{14} \cos \omega t\, y_4 + \\ &\quad + \lambda\,\omega\, h_{11} \sin \omega t\, y_1 + \lambda\,\omega\, h_{12} \sin \omega t\, y_2 - \lambda\,\omega\, d_{12}\, y_2 + \lambda\,\omega\, v_{11} \cos \omega t + \lambda\,\omega\, v_{21} \sin \omega t + n_1\,, \\ \dot y_3 &= \lambda\,\omega\, g_{31} \cos \omega t\, y_1 + \lambda\,\omega\, g_{33} \cos \omega t\, y_3 + \lambda\,\omega\, g_{34} \cos \omega t\, y_4 - \lambda\,\omega\, d_{34}\, y_4 + \lambda\,\omega\, v_{13} \cos \omega t + n_3\,, \end{aligned} \tag{2.22}$$

$$\begin{aligned} \ddot y_4 + \omega_0^2\, y_4 = &- \lambda\,\omega\,\omega_0\, g_{21} \cos \omega t\, y_1 - \lambda\,\omega\,\omega_0\, g_{22} \cos \omega t\, y_2 - \lambda\,\omega\,\omega_0\, g_{23} \cos \omega t\, y_3 - \\ &- \lambda\,\omega\,\omega_0\, g_{24} \cos \omega t\, y_4 + \lambda\,\omega\, g_{41} \cos \omega t\, \dot y_1 + \lambda\,\omega\, g_{43} \cos \omega t\, \dot y_3 + \\ &+ \lambda\,\omega\, g_{44} \cos \omega t\, \dot y_4 - \lambda\,\omega\,\omega_0\, h_{21} \sin \omega t\, y_1 - \lambda\,\omega\,\omega_0\, h_{22} \sin \omega t\, y_2 - \\ &- \lambda\,\omega^2\, g_{41} \sin \omega t\, y_1 - \lambda\,\omega^2\, g_{43} \sin \omega t\, y_3 - \lambda\,\omega^2\, g_{44} \sin \omega t\, y_4 + \\ &+ \lambda\,\omega\,\omega_0\, d_{22}\, y_2 - \lambda\,\omega\, d_{44}\, \dot y_4 - \lambda\,\omega\,\omega_0\, v_{12} \cos \omega t - \\ &- \lambda\,\omega\,\omega_0\, v_{22} \sin \omega t - \lambda\,\omega^2\, v_{14} \sin \omega t + \dot n_4 - \omega_0\, n_2\,. \end{aligned} \tag{2.23}$$

In such a system two kinds of resonance may occur in the first order approximation. Resonance due to nonlinear terms occurs for ω close to ω_0 and a parametric resonance may be expected for ω close to $2\,\omega_0$. For other ranges of ω the stability of the system will not be affected in first order approximation.

The solution of the system will be written under the form

$$y_i = a_i + \sum_{k=1} \varepsilon^k\, u_i^{(k)} \quad (i = 1, 3)\,, \qquad y_4 = a_4 \cos \psi + \sum_{k=1} \varepsilon^k\, u_4^{(k)}\,, \tag{2.24}$$

where $u_i^{(k)}$ $(i = 1, 3, 4)$ is an explicit function of the amplitudes, the phase and time also, the time derivative of the amplitudes and phases are expended in power series of ε:

$$\begin{aligned} \frac{da_i}{dt} &= \sum_{k=1} \varepsilon^k\, A_i^{(k)}\,, \qquad (i = 1, 3, 4) \\ \frac{d\psi}{dt} &= \omega_0 + \frac{d\theta}{dt}, \end{aligned} \tag{2.25}$$

with

$$\frac{d\theta}{dt} = \sum_k \varepsilon^k\, \eta^k + \sum_k \varepsilon^k\, B^{(k)}\,.$$

For ω close to ω_0, small forcing terms may give rise to large responses mainly when damping is light. Consequently, the zeroth order system will be taken as

$$\dot y_1 = \dot y_3 = 0 \tag{2.26}$$

and

$$\ddot y_4 + \omega_0^2\, y_4 = -\lambda\,\omega\,(d_{22} + d_{44})\, \dot y_4 - \lambda\,\omega\,\omega_0\, v_{12} \cos \omega t\,. \tag{2.27}$$

The equation (2.27) has the form

$$\ddot y_4 + 2\,\zeta\,\omega_0\, \dot y_4 + \omega_0^2\, y_4 = \omega_0^2\, a_0 \cos \omega t\,,$$

where defining ϱ as $\varrho = \omega/\omega_0$,

$$\zeta = \lambda\,\varrho\,\frac{d_{22} + d_{44}}{2} = \frac{\lambda\,\varrho}{2}\,(\xi_1 + \xi_2)\,, \qquad a_0 = -\lambda\,\varrho\, v_{12} = \lambda\,\varrho\,\frac{c\,b}{d}.$$

The free solution of (2.27) will decrease asymptotically to zero and then the zeroth order solution of (2.23), $y_4 = a_4 \cos \psi$, will be taken as the forced solution of (2.22),

$$y_4 = a_4 \cos \psi = a_4 \cos (\omega t + \varphi)\,, \tag{2.28}$$

where

$$a_4 = a_0\, [(1 - \varrho^2)^2 + 4\,\zeta^2\,\varrho^2]^{-1/2}\,, \qquad \tan \varphi = \frac{2\,\zeta\,\varrho}{1 - \varrho^2}.$$

With zero initial condition the zeroth order solution of (2.22) is simply given by

$$y_1 = a_1 \quad \text{and} \quad y_3 = a_3\,, \tag{2.29}$$

with $a_1 = a_3 = 0$.

Due to the large value of the ratio a_4/a_0 the nonlinear terms in the right hand side of equations (2.22) and (2.23) will be much larger than the periodic and forcing terms.

If the nonlinear terms are then of the order of the parameter ε, the other terms will be considered as terms of the order of a certain power of ε. They will be neglected if only a first order solution has to be determined.

Then for the first order approximation, the system (2.22), (2.23) will be equivalent to:

$$\dot{y}_1 = n_1\,, \qquad \dot{y}_3 = n_3\,, \tag{2.30}$$

$$\ddot{y}_4 + \lambda\,\omega\,(d_{22} + d_{44})\,\dot{y}_4 + \omega_0^2\,y_4 = -\lambda\,\omega\,\omega_0\,v_{12}\cos\omega t + \dot{n}_4 - \omega_0\,n_2$$

where from (2.21)

$$n_1 = \omega_0\,\frac{b}{d}\,y_3\,y_4 - \beta\,\frac{b}{d}\,y_4^2\,, \qquad n_2 = -\,\omega_0\,\frac{b}{d}\,y_3\,y_4 + \beta\,\frac{b}{d}\,y_4^2\,,$$

$$n_3 = \beta\,\frac{b}{d}\,y_2\,y_3 - \frac{\beta^2}{\omega_0}\,\frac{b}{d}\,y_2\,y_4\,, \qquad n_4 = \omega_0\,\frac{b}{d}\,y_2\,y_3 - \beta\,\frac{b}{d}\,y_2\,y_4\,,$$

and with same approximation:

$$y_2 = -\,\frac{\dot{y}_4}{\omega_0} \quad \text{and} \quad \dot{y}_2 = \omega_0\,y_4\,. \tag{2.31}$$

The time derivative of a_1 and a_3 will be given by the constant part of the expansion in FOURIER series of the right-hand sides of (2.30) for y_1, y_2, y_3, and y_4 given by (2.28), (2.29) and (2.31).

The equation for the time derivative of a_3 can be solved independently and it is seen that a_3 remains a constant equal to zero with our assumption on initial conditions.

Consequently, there are no terms leading to resonance for y_4 and the parameters a_4 and φ will be also keep there value given by the zeroth order approximation.

Only the time derivative of a_1 will be different from zero and in first order approximation

$$\frac{da_1}{dt} = -\,\frac{\beta}{2}\,\frac{b}{d}\,a_4^2 \tag{2.32}$$

with a_4 given in (2.28).

The drift rates are equal to constant part of the time derivative of the variables x_1 and x_2 and are directly related to y by the relation (2.17).

There is no drift for the variables x_2. This means that the angle θ between the inner and the outer gimbals remains a constant.

On the other hand there is an important drift for the variable x_1. The drift rate is equal to da_1/dt and is then given by (2.32). It must be noted that the drift rate is independent of θ_{10} and but depends on θ_{20}.

As the nutation frequency ω is also independent of θ_{10}, the drift will not affect the value of the drift rate which remains constant for a constant value of ω, in other words the gyro will remain in the same resonance situation while drifting.

It must be noted that the drift rate is equal to zero when the gyroscope is in canonical position and vibrations will not affect the behavior of two degrees of freedom mounted on inertial platforms.

For ω close to $2\,\omega_0$, the forcing terms do not produce large responses and their influence will of the some order as the influence of parametric and nonlinear terms.

Consequently, the zeroth order system will be taken as

$$\dot{y}_1 = \dot{y}_3 = 0 \qquad \text{and} \qquad \ddot{y}_4 = \omega_0^2\,y_4 = 0\,. \tag{2.33}$$

The solution of this system will then be

$$y_1 = a_1\,, \qquad y_3 = a_3\,, \qquad y_4 = a_4\cos\psi = a_4\cos(\omega_0\,t + \varphi)\,. \tag{2.34}$$

In the resonance region it is convenient to write $\omega\,t = 2\,\psi - 2\,\theta$ and the last equation (2.34) can be written

$$y_4 = a_4\cos\left(\frac{\omega\,t}{2} + \theta\right). \tag{2.35}$$

If we assume that initial condition on a_4 is of the order of the amplitude λ and considering (2.31), the system (2.22) will be equivalent up to the first order approximation, to the system

$$\dot{y}_1 = n_1\,, \qquad \dot{y}_3 = n_3\,,$$

$$\ddot{y}_4 + \omega_0^2\,y_4 = -\,\lambda\,\omega\,(d_{22} + d_{44})\,\dot{y}_4 + \lambda\,\omega\,h_{22}\sin\omega\,t\,\dot{y}_4 - \lambda\,\frac{\omega^2}{2}\,g_{24}\cos\omega\,t\,y_4 - \lambda\,\omega^2\,g_{44}\sin\omega\,t\,y_4 +$$

$$+\,\lambda\,\omega\,(g_{44} + g_{22})\cos\omega\,t\,\dot{y}_4\,. \tag{2.36}$$

Substituing the various parameters by their values given by (2.21) and noting that $g_{22} = - g_{44}$, the last equation (2.36) may be written:

$$\ddot{y}_4 + \omega_0^2 y_4 = - \lambda \omega (\xi_1 + \xi_2) \dot{y}_4 + \lambda \omega \frac{c \omega}{\beta} \sin \omega t \dot{y}_4 - \lambda \frac{\omega^2}{2} \left(\frac{2 \beta}{\omega} - \frac{\omega}{2 \beta} \right) \cos \omega t y_4 +$$

$$+ \lambda \omega^2 \frac{a b}{d} \sin \omega t y_4 . \tag{2.37}$$

This equation is a rather simple HILL's equation which may be solved exactly without difficulty. The parameter λ being considered as a small quantity equal to ε, the method for slightly Nonlinear Dynamic Systems can be used to solve the problem and it will be checked that the solution obtained by this method is the exact solution. As a result of the method, the parameters $A_4^{(1)}$ and $B^{(1)}$ in equations (2.25) are given by:

$$A_4^{(1)} = - \frac{\bar{f}_{41}^{s*}}{\omega} , \qquad B^{(1)} = - \frac{\bar{f}_{41}^{c*}}{a_4 \omega} , \tag{2.38}$$

where $\bar{f}_{41}^{s*}$ and $\bar{f}_{41}^{c*}$ are the coefficients of $\sin \psi$ and $\cos \psi$ in the expansion in FOURIER series of the right hand sides of (2.37), after substitution of y_4 and $\dot{y}_4$ by their zeroth order value, i.e.:

$$y_4 = a_4 \cos \psi \quad \text{and} \quad \dot{y}_4 = - a_4 \frac{\omega}{2} \sin \psi . \tag{2.39}$$

This system (2.25) is then equal upto the first order to

$$\frac{d a_4}{d t} = - C + A \sin 2 \theta + B \cos 2 \theta \, a_4 , \qquad \frac{d \theta}{d t} = \eta + A \cos 2 \theta - B \sin 2 \theta , \tag{2.40}$$

where

$$A = \frac{\lambda \omega}{2} \left(\frac{\beta}{\omega} + \frac{\omega}{4 \beta} \right) c , \qquad B = \frac{\lambda \omega}{2} \frac{a b}{d} , \qquad C = \lambda \frac{\omega}{2} (\xi_1 + \xi_2) = \frac{1}{2} \left(\frac{f_1}{I} + \frac{f_2}{I_2} \right) .$$

After the change of variable

$$u = a_4 \cos \theta , \qquad v = a_4 \sin \theta , \tag{2.41}$$

a linear system with constant coefficients is obtained for the variable u and v

$$\frac{d u}{d t} = (- C + B) u + (- \eta + A) v , \qquad \frac{d v}{d t} = (\eta + A) u - (C + B) v . \tag{2.42}$$

The characteristic equation of this system is

$$m^2 + 2 C m + (\eta^2 + C^2 - A^2 - B^2) = 0 . \tag{2.43}$$

The system (2.42) and consequently (2.37), is stable when

$$C \geqq 0 \tag{2.44}$$

and

$$\eta^2 \geqq A^2 + B^2 - C^2 . \tag{2.45}$$

It must be noted that the characteristic equation is equal to the first order characteristic equation of the exact solution given from FLOQUET's theory. For a general discussion of this problem we refer to [10].

The condition (2.44) is always satisfied for systems with physical damping. The condition (2.45) will be satisfied for all the values of ω, if

$$C^2 > A^2 + B^2 .$$

The parameters A and B being functions of the amplitude λ, the condition (2.45) will always be satisfied if the vibration amplitude remain smaller than the critical value

$$\lambda = \frac{1}{\omega} \left(\frac{f_1}{I} + \frac{f_2}{I_2} \right) \left[\left(\frac{\beta}{\omega} + \frac{\omega}{4 \beta} \right)^2 + \frac{a^2 b^2}{d^2} \right]^{-1/2}$$

for ω in the neighborhood of $2 \omega_0$.

Other parametric resonances may occur for the system but one must look at higher order approximation and due to the presence of damping such resonance will not be obtained for realistic vibration amplitudes and, further, the characteristic exponents will very small and these resonances will be then considered as secondary effects.

From this part of analysis it is concluded that better gyroscopes, as far as friction is concerned, must be protected more efficiently against vibrations in order to avoid parametric instability.

Knowing the solution (2.42) y_4 may be obtained from (2.35) as

$$y_4 = a_4\left(\cos\frac{\omega t}{2} + \theta\right) = \cos\frac{\omega t}{2}\, u - \sin\frac{\omega t}{2}\, v\,.$$

The drift will be given as previously by (2.32) and it will remain small, if not negligible, outside the resonance region, but will take an inadmissible value when parametric resonance occurs.

It must be noted that the drift rate will, in this case, be equal to zero when the gyroscope is in canonical position. But the value of a_4 may take infinite value at least theoritically. In practice obsviously a_4 will remain bounded but the gyroscope may fail before that.

The possibility of such a resonance was pointed out by YANSOUNI in an unpublisher paper [11] but the evidence of this resonance could not be obtained at that time. An experimental verification of the results of this paper was performed in the laboratories of the Department of Applied Mechanics at the University of Louvain and the authors wish to express their gratitude to M. COLIN and MEYERS for their help.

Conclusions

This paper present a complete description of the kinematical drift of mechanically perfect two degrees of freedom gyroscopes subject to monofrequency vibration about some axis perpendicular to the outer gimbal axis. The equations of motion are slightly nonlinear equations, with time-varying parameters, when the amplitude of the external vibration is small. This assumption is quite realistic for strap-down instruments in vehicules. The most important drift occurs when the vibration frequency is close to the natural nutation frequency of the gyroscope. This nonlinear drift is well-known and we reformulated the problem and just presented the solution in a more rigorous form.

Besides this drift, a first order parametric resonance may also occur when the vibration frequency is almost equal to twice the nutation frequency. Gyroscopes with very small damping coefficients will be subject to this kind of resonance for rather small vibrations amplitudes and should need a better protection.

References

1 DE JONQUIERES, E., Au sujet de certaines circonstances qui se présentent dans le mouvement de la toupie, Comptes Rendus Acad. de Sc. de Paris **102** (1886).

2 CRAWLEY, J. and MAUNDER, L., Drift of a Vibrating Gyroscope supported in Gimbals, Proc. of the 11th Congr. of Appl. Mech., Munich, 1964.

3 MAGNUS, K., Beiträge zur Dynamik des kräftefreien, kardanisch gelagerten Kreisels, ZAMM **35** (1955).

4 PLYMATE, B. T. and GOODSTEIN, E., Nutation of a free Gyro subjected to an Impulse, J. of Appl. Mech. 22,3, p. 365 (1955).

5 MAUNDER, L., Aspect of the Gimbal Suspension of Gyroscopes, Gyrodynamics, (Ed. H. Ziegler), IUTAM symposium, Celerina, 1962.

6 LEMAITRE, PH. and VAN VINCKENROY, A., Drift compensation in two-axes gyroscopes, Seminaire de Math. Appl. et Méc., Univ. de Louvain, rapport no 29, sept. 1969.

7 Климов, Д. М., Слеюкин, Л. Н., Применение асимтотических методов к решение задач о движении астатического гироскопа в кардановом подвесе, Известия Академии Наук СССР, Отн. Механика и Машиностроение, No 3, 1963.

8 Климов, Д. М., О Движении гироскопа в кардановом подвесе с неаксиально насаженным ротором, Докл. Акад. Начк. СССР (1959), T. **124**, No 3, also transl. as Soviet Phys. Dokl. **124**, no 4, 1959.

9 WENGLARZ, R. A. and KANE, T. R., Gyroscopic Drifts associated with rotational Vehicle Vibrations, J. of Appl. Mech. **35**, p. 553 (1968).

10 NISHIKAWA, Y. and WILLEMS, P. Y., A Method for Stability Investigation of a periodic dynamic System with many Degrees of Freedom, J. of the Franklin Inst., Vol. **287**, no 2, feb. 1969.

11 YANSOUNI, P., Dérive d'un gyroscope soumis à bord d'un véhicule à des perturbations aléatoires, Seminaire de Math. Appl. et Mec. Univ. de Louvain, Rapport no 13, Déc. 66.

Eingereicht am 3. 4. 1973

Anschriften: P. Y. WILLEMS, Université Catholique de Louvain, Unité de Mécanique Appliquée, Bâtiment Simon Stévin, Place du Levant 3, B-1348 Louvain-la-Neuve, Belgique;
Y. NISHIKAWA, Kyoto Univ., Dept. of Electrical Engineering, Yoshida, Kyoto 606, Japan.

ZAMM 54, 181 – 191 (1974)

L. I. WEINGARTEN *) / H. REISMANN **)

Forced Motion of Cylindrical Shells — A Comparison of Shell Theory with Elasticity Theory [1])

Die dynamische Reaktion einer frei gelagerten kreiszylindrischen Schale wird im Rahmen der dreidimensionalen Elastizitätstheorie behandelt. Für eine spezielle Lastverteilung und Schalengeometrie wird die so gewonnene Lösung mit derjenigen, die man im Rahmen der Flüggeschen und der „verbesserten" Schalentheorie erhält, numerisch verglichen. Es zeigt sich, daß — gemessen an der elastizitätstheoretischen Lösung — die „verbesserte" Theorie hinsichtlich der Voraussage des Reaktionsverhaltens der Schale über Zeiten, die die Spannungs- und Verschiebungsmaxima enthalten, bessere Ergebnisse liefert als die Flüggesche Theorie und daß außerdem Vorsicht am Platze ist, wenn eine der Schalentheorien auf das anfängliche Reaktionsverhalten angewendet wird.

The dynamic response of a freely supported circular cylindrical shell is treated within the framework of the three dimensional theory of elasticity. A numerical comparison of this solution with corresponding solutions within the framework of the Flügge and "improved" shell theories is presented for a specific load distribution and shell geometry. When compared to the elasticity theory solution, it is shown that improved theory is better than Flügge theory in predicting the response over times which contain the maximum stresses and displacements. The results also show that caution must be used when applying either of the shell theories to the initial response.

В рамках трёхмерной теории упругости рассматривается динамическая реакция покоящейся свободно оболочки круговой цилиндрической формы. Для определённого распределения нагрузки и геометрии оболочки производится численное сравнение полученных решений, которые вытекают из теории оболочка Флюгге и из „улучшенной" теории оболочек. Оказывается, что по сравнению с упруго-теоретическим решением „улучшенная" теория даёт относительно предсказания о поведении реакции оболочки по времени, содержащем максимумы напряжения и смещения, лучшие результаты чем теория Флюгге. При применении одной из теорий оболочек для рассмотра начального поведения реакций, следует соблюдать особую осторожность.

Nomenclature (I)

Dimensional Quantity	Physical Description	$F - L - T$ Units
λ, μ	LAMÉ's constants	F/L^2
ϱ	material density	$F\,T^2/L^4$
$c_1 = \sqrt{\dfrac{\lambda + 2\mu}{\varrho}}$	dilational wave velocity	L/T
$c_2 = \sqrt{\dfrac{\mu}{\varrho}}$	shear wave velocity	L/T
R	median surface radius	L
h	shell thickness	L
ν	POISSON's ratio	1
L	length of shell	L
$k^2 = \dfrac{c_1^2}{c_2^2} = \dfrac{\lambda + 2\mu}{\mu} = \dfrac{2(1-\nu)}{1-2\nu}$	wave speed ratio squared	1

Nomenclature (II)

Dimensionless Quantity	To convert to dimensional form multiply by	Physical Description
$\gamma = L/R$	1	length to radius ratio
$\eta = h/R$	1	thickness ratio
b	R	outer surface radius
a	R	inner surface radius
$a \leqq r \leqq b$	R	radial coordinate
$0 \leqq \varphi \leqq 2\pi$	1	plane polar angle
$0 \leqq z \leqq \gamma$	R	axial coordinate
$0 \leqq t < \infty$	R/c_2	time
Ω	c_2/R	frequency
u_r, u_φ, u_z	R	radial, circumferential and axial displacements, respectively
$\tau_{rr}, \tau_{\varphi\varphi}, \tau_{zz}$	ϱc_2^2	radial, circumferential and axial stresses
$\tau_{r\varphi}, \tau_{rz}, \tau_{\varphi z}$	ϱc_2^2	shear stresses
$\varepsilon_{rr}, \varepsilon_{\varphi\varphi}, \varepsilon_{zz}$	1	radial, circumferential and axial strains
$\varepsilon_{r\varphi}, \varepsilon_{rz}, \varepsilon_{\varphi z}$	1	shear strains
$\Phi, \Psi_r, \Psi_\varphi, \Psi_z$	R^2	displacement potentials
$\boldsymbol{n}$		
$\boldsymbol{T}$	ϱc_2^2	traction vector acting on S with outer unit normal n
P_0	ϱc_2^2	radial load intensity

I. Introduction

When determining the dynamic response of a finite cylindrical shell, the analyst has at his disposal several shell theories. The mathematical complexity of these theories is directly related to the accuracy with which they characterize the actual motion of the shell. When considering an elastic material with sufficiently small deformations, the standard of accuracy is the classical equations of elasticity [1].

*) Staff Member, Applied Mechanics Division.

**) Professor, Faculty of Engineering and Applied Sciences.

[1]) This work was supported through U. S. Air Force, AFOSR, Grant No. 71-1971 and in part by the U. S. Atomic Energy Commission. Computer facilities were generously made available to the authors by the Computing Center at the State University of New York at Buffalo, which is partially supported by N. I. H. Grant FR-00126 and NSF Grant GP-7318.

Due to the analytical difficulties involved in determining the response of cylinders via the equations of the classical theory of elasticity, few solutions of problems using this theory exist. For most meaningful problems, the mathematical complexity involved in finding a quantitative description of the response requires the use of high speed computers. For these reasons, most of the papers utilizing this theory to analyze finite cylindrical shells have appeared recently. References [2] through [15] contain relevant publications on the response of finite cylinders within the classical theory of elasticity.

With the knowledge that solutions within the framework of the three dimensional theory of elasticity are mathematically, complex, several shell theories were developed. The simplest, membrane theory, permits no variation of the displacements or stresses through the shell thickness. Membrane theory equations for cylindrical shells are found in [16] and [17]. Theories developed by Love [18], Flügge [16] and Vlasov [17] include bending effects in addition to membrane effects. This is done by permitting the dependent variables to change through the shell thickness such that straight lines initially normal to the median surface remain normal during deformation. This is comparable to the Euler-Bernoulli equations in beam theory. A theory which is sophisticated in the manner in which it characterizes shell behavior is commonly denoted as the improved theory. This theory includes the effects of shear deformation and rotatory inertia. Although the line elements, initially normal to the median surface, still remain straight and retain their original length, a further improvement is obtained by introducing a correction factor into the transverse shear force to compensate for this. The numerical value of this factor is obtained by matching the phase velocity of the lowest mode of propagation of free harmonic waves with that obtained from the three dimensional equations of elasticity [6]. The improved theory equations are found in [19] and [20].

In order to determine the accuracy of response predicted by shell theories, we compare them to those developed by the three dimensional theory of elasticity. No full comparison, with numerical results, has yet been made for the static and dynamic response of the freely supported hollow cylindrical shell to a suddenly applied load. This type of comparison has been made for a cylindrical shell in plane strain by Pawlik and Reismann [21]. Since most engineering structures, like the one described above, are analyzed by shell theories, it is of interest to know the relative accuracy of these theories as compared to elasticity theory. It is the purpose of this paper to present an investigation which does compare the static and dynamic responses of such a shell.

In this problem, the finite cylindrical shell of circular cross section is suddenly loaded by a radially inward directed load applied to its outer surface. The end surfaces of the shell are freely supported. The distribution of this load is symmetric with respect to the polar angle, but otherwise arbitrary. The forced motion response is then calculated by the three-dimensional theory of elasticity. The response predicted by this theory for particular shell parameters and load distribution is then compared to results obtained from Flügge and improved shell theories.

II. Statement of the boundary value problem

A linearly elastic, isotropic, and homogeneous finite closed cylindrical shell is at rest and in an unstrained state. The shell is freely supported at its end surfaces. A load, directed radially inward, is suddenly applied to its outer cylindrical surface. The distribution of this load is symmetric about the plane $\varphi = 0$, but otherwise arbitrary. The response to a completely general load distribution can be found by the superposition principle. No body forces are assumed to be acting. The initial geometry of the cylindrical shell is shown in Figure 1.

The load can be represented as follows:

$$P(\varphi, z, t) = P_0 \, f(\varphi) \, g(z) \, H(t) \, . \tag{1}$$

In (1), P_0 is a reference pressure, $f(\varphi)$ is the load distribution along a constant z plane, $g(z)$ is the load distribution along a constant φ plane, and $H(t)$ is the Heaviside step function with respect to time.

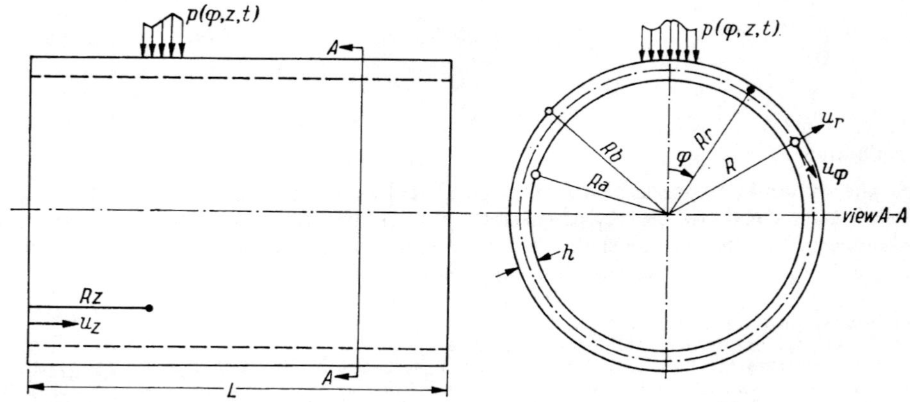

Fig. 1. Shell geometry

For a load distribution as described above, $f(\varphi)$ and $g(z)$ are expressed in terms of a FOURIER series

$$f(\varphi) = \frac{1}{\pi}\left[\frac{a_0}{2} + \sum_{n=1}^{\infty} \bar{a}_n \cos n\,\varphi\right], \qquad g(z) = \frac{1}{\pi}\sum_{m=1}^{\infty} \bar{b}_m \sin\frac{m\pi}{\gamma} z\,, \tag{2}$$

where $\gamma = L/R$.

In determining the response to an actual loading, these summations will have to be truncated after summing to $n = N - 1$, and $m = M - 1$. Hence, we approximate $f(\varphi)$ and $g(z)$ by $f_{N-1}(\varphi)$ and $g_{M-1}(z)$, respectively. The accuracy of these approximations can be improved utilizing LANCZOS' smoothing technique, as explained in references [21], [22] and [23]. This amounts to multiplying each term of $f_{N-1}(\varphi)$ by $\left(\frac{N}{n\pi}\sin\frac{n\pi}{N}\right)$ and $g_{M-1}(z)$ by $\left(\frac{M}{m\pi}\sin\frac{m\pi}{M}\right)$;

$$f_{N-1}(\varphi) = \frac{1}{\pi}\left[\frac{a_0}{2} + \sum_{n=1}^{N-1} a_n \cos n\,\varphi\right], \qquad g_{M-1}(z) = \frac{1}{\pi}\sum_{m=1}^{M-1} b_m \sin\frac{m\pi}{\gamma} z\,, \tag{3}$$

where

$$a_n = \bar{a}_n\left(\frac{N}{n\pi}\sin\frac{n\pi}{N}\right), \qquad b_m = \bar{b}_m\left(\frac{M}{m\pi}\sin\frac{m\pi}{M}\right).$$

Therefore, the actual loading to be considered is as follows:

$$P(\varphi, z, t) = P_0\, f_{N-1}(\varphi)\, g_{M-1}(z)\, H(t)\,. \tag{4}$$

For convenience, all variables used in the analysis are in dimensionless form. Multipliers given in the Nomenclature can be used to convert to dimensional form.

III. Theory of elasticity

A. Basic equations and method of solution

The motion of a linearly elastic, isotropic and homogeneous material restricted to sufficiently small deformation is characterized by NAVIER's equations:

$$\nabla^2 \boldsymbol{u} + \frac{1}{1 - 2\nu}\nabla(\nabla\cdot\boldsymbol{u}) = \frac{\partial^2 \boldsymbol{u}}{\partial t^2}, \tag{5}$$

where $\boldsymbol{u} = u_r\,\boldsymbol{e}_r + u_\varphi\,\boldsymbol{e}_\varphi + u_z\,\boldsymbol{e}_z$ and $\boldsymbol{e}_r, \boldsymbol{e}_\varphi, \boldsymbol{e}_z$ are the unit vectors in the radial, circumferential, and axial directions respectively. The kinematic and constitutive relations are:

$$\varepsilon_{rr} = \frac{\partial u_r}{\partial r}, \qquad \varepsilon_{r\varphi} = \frac{1}{2}\left(\frac{1}{r}\frac{\partial u_r}{\partial\varphi} + \frac{\partial u_\varphi}{\partial r} - \frac{u_\varphi}{r}\right), \quad \text{etc.} \tag{6}$$

$$\sigma_{rr} = (k^2 - 2)\left(\frac{\partial u_r}{\partial r} + \frac{u_r}{r} + \frac{1}{r}\frac{\partial u_\varphi}{\partial\varphi} + \frac{\partial u_z}{\partial z}\right) + 2\frac{\partial u_r}{\partial r}, \qquad \sigma_{r\varphi} = \frac{1}{r}\frac{\partial u_r}{\partial\varphi} + \frac{\partial u_\varphi}{\partial r} - \frac{u_\varphi}{r}, \quad \text{etc.} \tag{7}$$

where $\varepsilon_{rr}, \varepsilon_{\varphi\varphi}, \varepsilon_{zz}, \sigma_{rr}, \sigma_{\varphi\varphi}, \sigma_{zz}$ are the normal strains and corresponding stresses, and $\varepsilon_{r\varphi}, \varepsilon_{rz}, \varepsilon_{\varphi z}, \sigma_{r\varphi}, \sigma_{rz}, \sigma_{\varphi z}$ are the shear angles and corresponding stresses.

The stress vector acting on the boundary surface S of the cylindrical shell is given by

$$\begin{aligned}
\overset{n}{\boldsymbol{T}} &= \overset{n}{T}_r\,\boldsymbol{e}_r + \overset{n}{T}_\varphi\,\boldsymbol{e}_\varphi + \overset{n}{T}_z\,\boldsymbol{e}_z\,,\\
\overset{n}{T}_r &= \tau_{rr}\,a_{nr} + \tau_{r\varphi}\,a_{n\varphi} + \tau_{rz}\,a_{nz}\,,\\
\overset{n}{T}_\varphi &= \tau_{r\varphi}\,a_{nr} + \tau_{\varphi\varphi}\,a_{n\varphi} + \tau_{z\varphi}\,a_{nz}\,,\\
\overset{n}{T}_z &= \tau_{rz}\,a_{nr} + \tau_{\varphi z}\,a_{n\varphi} + \tau_{zz}\,a_{nz}\,,
\end{aligned} \tag{8}$$

where $\hat{n}$ is the unit exterior normal to S and $a_{nr}, a_{n\varphi}, a_{nz}$ are its direction cosines.

A properly posed problem in elastokinetics may be formulated as follows: Find the displacement vector $\boldsymbol{u}(r, \varphi, z, t)$ in V (the cylindrical shell bounded by S) satisfying (5) for $t > 0$ for specified boundary conditions, i.e., if

$$\boldsymbol{u}(r, \varphi, z, t) = \boldsymbol{f}(r, \varphi, z, t) = f_r\,\boldsymbol{e}_r + f_\varphi\,\boldsymbol{e}_\varphi + f_z\,\boldsymbol{e}_z\,, \tag{9a}$$

$$\overset{n}{\boldsymbol{T}} = \boldsymbol{g}(r, \varphi, z, t) = g_r\,\boldsymbol{e}_r + g_\varphi\,\boldsymbol{e}_\varphi + g_z\,\boldsymbol{e}_z\,, \tag{9b}$$

on the surface S, then in every sub-region of the surface S, one member of each of the three pairs of functions (f_r, g_r), (f_φ, g_φ), and (f_z, g_z) are specified. In addition the following initial conditions are imposed in V:

$$\boldsymbol{u}(r, \varphi, z, 0) = \boldsymbol{u}^{(0)}(r, \varphi, z) , \qquad \frac{\partial \boldsymbol{u}(r, \varphi, z, 0)}{\partial t} = \frac{\partial \boldsymbol{u}^{(0)}(r, \varphi, z)}{\partial t} . \tag{10}$$

A method of solving the problem characterized by (5) through (10) is derived in [24]. This method is commonly known as the method of WILLIAMS or mode-acceleration method. The results of this reference, as they apply to this problem, are now summarized.

The solution of (5) is given by

$$\boldsymbol{u}(r, \varphi, z, t) = \boldsymbol{u}^{(s)}(r, \varphi, z, t) + \sum_{p=1}^{\infty} \boldsymbol{U}^{(p)}(r, \varphi, z)\, q^{(p)}(t) . \tag{11}$$

The *static* solution satisfies the equations

$$\nabla^2 \boldsymbol{u}^{(s)} + \frac{1}{1-2\nu} \nabla (\nabla \cdot \boldsymbol{u}^{(s)}) = 0 \tag{12}$$

and the associated boundary conditions

$$\boldsymbol{u}^{(s)} = \boldsymbol{f}(r, \varphi, z, t) , \qquad \overset{n}{\boldsymbol{T}}{}^{(s)} = \boldsymbol{g}(r, \varphi, z, t) \tag{13}$$

on S, where

$$\overset{n}{\boldsymbol{T}}{}^{(s)} = \overset{n}{T}{}^{(s)}_r \boldsymbol{e}_r + \overset{n}{T}{}^{(s)}_\varphi \boldsymbol{e}_\varphi + \overset{n}{T}{}^{(s)}_z \boldsymbol{e}_z , \tag{14}$$

$$\overset{n}{T}{}^{(s)}_r = \tau^{(s)}_{rr} a_{nr} + \tau^{(s)}_{r\varphi} a_{n\varphi} + \tau^{(s)}_{rz} a_{nz} , \quad \text{etc.}$$

$$\varepsilon^{(s)}_{rr} = \frac{\partial u^{(s)}_r}{\partial r} , \quad \text{etc.} \tag{15}$$

$$\tau^{(s)}_{rr} = (k^2 - 2)\left(\frac{\partial u^{(s)}_r}{\partial r} + \frac{u^{(s)}_r}{r} + \frac{1}{r}\frac{\partial u^{(s)}_\varphi}{\partial \varphi} + \frac{\partial u^{(s)}_z}{\partial z}\right) + 2\frac{\partial u^{(s)}_r}{\partial r} , \quad \text{etc.} \tag{16}$$

Only one member of each of the three pairs of functions (f_r, g_r), (f_φ, g_φ), and (f_z, g_z) appearing in (13) may be specified in any sub-region of the surface S.

The eigenfunctions are given by

$$\boldsymbol{U}^{(p)}(r, \varphi, z) = U^{(p)}_r(r, \varphi, z)\, \boldsymbol{e}_r + U^{(p)}_\varphi(r, \varphi, z)\, \boldsymbol{e}_\varphi + U^{(p)}_z(r, \varphi, z)\, \boldsymbol{e}_z . \tag{17}$$

The eigenfunctions and associated frequencies, Ω_p, satisfy the equations

$$\nabla^2 \boldsymbol{U}^{(p)} + \frac{1}{1-2\nu} \nabla (\nabla \cdot \boldsymbol{U}^{(p)}) = -\Omega_p^2 \boldsymbol{U}^{(p)} , \tag{18}$$

subject to the homogeneous boundary conditions (on S)

$$U^{(p)}_r(r, \varphi, z) = 0 \quad \text{if} \quad f_r(r, \varphi, z, t) \quad \text{is specified} , \tag{19a}$$

$$U^{(p)}_\varphi(r, \varphi, z) = 0 \quad \text{if} \quad f_\varphi(r, \varphi, z, t) \quad \text{is specified} , \tag{19b}$$

$$U^{(p)}_z(r, \varphi, z) = 0 \quad \text{if} \quad f_z(r, \varphi, z, t) \quad \text{is specified} , \tag{19c}$$

$$\overset{n}{T}{}^{(p)}_r = \tau^{(p)}_{rr} a_{nr} + \tau^{(p)}_{r\varphi} a_{n\varphi} + \tau^{(p)}_{rz} a_{nz} = 0 \quad \text{if} \quad g_r(r, \varphi, z, t) \quad \text{is specified} , \tag{19d}$$

$$\overset{n}{T}{}^{(p)}_\varphi = \tau^{(p)}_{r\varphi} a_{nr} + \tau^{(p)}_{\varphi\varphi} a_{n\varphi} + \tau^{(p)}_{\varphi z} a_{nz} = 0 \quad \text{if} \quad g_\varphi(r, \varphi, z, t) \quad \text{is specified} , \tag{19e}$$

$$\overset{n}{T}{}^{(p)}_z = \tau^{(p)}_{rz} a_{nr} + \tau^{(p)}_{\varphi z} a_{n\varphi} + \tau^{(p)}_{zz} a_{nz} = 0 \quad \text{if} \quad g_z(r, \varphi, z, t) \quad \text{is specified} , \tag{19f}$$

where $\overset{n}{\boldsymbol{T}}{}^{(p)} = \overset{n}{T}{}^{(p)}_r \boldsymbol{e}_r + \overset{n}{T}{}^{(p)}_\varphi \boldsymbol{e}_\varphi + \overset{n}{T}{}^{(p)}_z \boldsymbol{e}_z$.

The strains and stresses due to the modal displacements are given by

$$\varepsilon^{(p)}_{rr} = \frac{\partial U^{(p)}_r}{\partial r} , \quad \text{etc.} \tag{20}$$

$$\tau^{(p)}_{rr} = (k^2 - 2)\left(\frac{\partial U^{(p)}_r}{\partial r} + \frac{U^{(p)}_r}{r} + \frac{1}{r}\frac{\partial U^{(p)}_\varphi}{\partial \varphi} + \frac{\partial U^{(p)}_z}{\partial z}\right) + 2\frac{\partial U^{(p)}_r}{\partial r} , \quad \text{etc.} \tag{21}$$

Equations (18) and (19) specify each normal mode only to within an indeterminate constant multiplier common to each member of the set of three functions $U^{(p)}_r$, $U^{(p)}_\varphi$, and $U^{(p)}_z$. By introducing a mode normalization condition, unique expressions for the mode shapes are obtained. In the present case the form of this condition is selected because of its convenience in the subsequent solution of the forced motion problem and it is given by

$$\int_V [U^{(p)}_r U^{(p)}_r + U^{(p)}_\varphi U^{(p)}_\varphi + U^{(p)}_z U^{(p)}_z]\, dV = 1 . \tag{22}$$

The time dependent function $q^{(p)}(t)$ in (11) is given by

$$q^{(p)}(t) = q^{(p)}(0) \cos \Omega_p t + \frac{1}{\Omega_p} \dot{q}^{(p)} q(0) \sin \Omega_p t + \frac{1}{\Omega_p} \int_0^t \ddot{Q}^{(p)}(t_1) \sin \Omega_p (t - t_1)\, dt_1 , \tag{23}$$

where

$$q^{(p)}(0) = \int_V [u_r^{(0)} U_r^{(p)} + u_\varphi^{(0)} U_\varphi^{(p)} + u_z^{(0)} U_z^{(p)}]\, dV + Q^p(0) , \tag{24}$$

$$\dot{q}^{(p)}(0) = \int_V [\dot{u}_r^{(0)} U_r^{(p)} + \dot{u}_\varphi^{(0)} U_\varphi^{(p)} + \dot{u}_z^{(0)} U_z^{(p)}]\, dV + \dot{Q}(0) \tag{25}$$

and

$$\Omega_p^2\, Q^{(p)}(t) = \int_S \left[\overset{n}{T}{}_r^{(p)} f_r + \overset{n}{T}{}_\varphi^{(p)} f_\varphi + \overset{n}{T}{}_z^{(p)} f_z\right] dS - \int_S [U_r^{(p)} g_r + U_\varphi^{(p)} g_\varphi - U_z^{(p)} g_z]\, dS . \tag{26}$$

For the present investigation as illustrated in Section II, we must find the response $\boldsymbol{u}$ that satisfies (5), subject to the following boundary and initial conditions.

(a) Boundary conditions:

(i) freely supported edges

$$u_r(r, \varphi, 0, t) = u_\varphi(r, \varphi, 0, t) = \tau_{zz}(r, \varphi, 0, t) = 0 , \tag{27a}$$

$$u_r(r, \varphi, \gamma, t) = u_\varphi(r, \varphi, \gamma, t) = \tau_{zz}(r, \varphi, \gamma, t) = 0 , \tag{27b}$$

(ii) tractions on inner and outer surfaces

$$\tau_{rz}(a, \varphi, z, t) = \tau_{r\varphi}(a, \varphi, z, t) = \tau_{rr}(a, \varphi, z, t) = 0 , \tag{28a}$$

$$\tau_{rz}(b, \varphi, z, t) = \tau_{r\varphi}(b, \varphi, z, t) = 0 , \tag{28b}$$

$$\tau_{rr}(b, \varphi, z, t) = - P_0 f_{N-1}(\varphi)\, g_{M-1}(z)\, H(t) ,$$

where $a = 1 - \frac{\eta}{2}$ and $b = 1 + \frac{\eta}{2}$, $\eta = \frac{h}{R}$.

(b) initial conditions:

$$\boldsymbol{u}(r, \varphi, z, 0) = 0 , \qquad \frac{\partial \boldsymbol{u}(r, \varphi, z, 0)}{\partial t} = 0 . \tag{29}$$

B. The static solution

Equations (12) are now solved subject to the boundary conditions (27) and (28) where it is understood that u_r, u_φ, τ_{zz}, τ_{rz}, $\tau_{r\varphi}$, and τ_{rr} are replaced by $u_r^{(s)}$, $u_\varphi^{(s)}$, $\tau_{zz}^{(s)}$, $\tau_{rz}^{(s)}$, $\tau_{r\varphi}^{(s)}$, and $\tau_{rr}^{(s)}$, respectively. Using the NEUBER-PAPCOVITCH three-function theorem, found in references [25] and [26], the displacements are given by

$$u_r^{(s)} = \Psi_1 \cos \varphi + \Psi_2 \sin \varphi - \frac{1}{4(1-\nu)} \frac{\partial}{\partial r} (\Psi_1 r \cos \varphi + \Psi_2 r \sin \varphi + \Psi_4 + z \Psi_3) , \tag{30a}$$

$$u_\varphi^{(s)} = - \Psi_1 \sin \varphi + \Psi_2 \cos \varphi - \frac{1}{4(1-\nu)} \frac{1}{r} \frac{\partial}{\partial \varphi} (\Psi_1 r \cos \varphi + \Psi_2 r \sin \varphi + z \Psi_3 + \Psi_4) , \tag{30b}$$

$$u_r^{(s)} = - \frac{1}{4(1-\nu)} \frac{\partial}{\partial z} (\Psi_1 r \cos \varphi + \Psi_2 r \sin \varphi + \Psi_4 + z \Psi_3) , \tag{30c}$$

where Ψ_i satisfies

$$\nabla^2 \Psi_i = 0 \qquad (i = 1, 2, 3, 4) . \tag{31}$$

Since the loading is periodic,

$$\Psi_3 = 0 . \tag{32}$$

Three harmonic functions Ψ_1, Ψ_2, and Ψ_4 must be found, together with six coefficients to satisfy the boundary conditions for each term of the FOURIER series. The functions that satisfy these requirements are

$$\Psi_1 = \sum_{n=1}^{N-1} \sum_{m=1}^{M-1} \left[(A_{mn} + B_{mn})\, I_n\!\left(\frac{m\pi}{\gamma} r\right) + (E_{mn} + F_{mn})\, K_n\!\left(\frac{m\pi}{\gamma} r\right)\right] \cos n\varphi \sin \frac{m\pi}{\gamma} z , \tag{33a}$$

$$\Psi_2 = \sum_{n=1}^{N-1} \sum_{m=1}^{M-1} \left[(A_{mn} + B_{mn})\, I_n\!\left(\frac{m\pi}{\gamma} r\right) + (E_{mn} - F_{mn})\, K_n\!\left(\frac{m\pi}{\gamma} r\right)\right] \sin n\varphi \sin \frac{m\pi}{\gamma} z , \tag{33b}$$

$$\Psi_4 = \sum_{n=1}^{N-1} \sum_{m=1}^{M-1} \left[C_{mn}\, I_n\!\left(\frac{m\pi}{\gamma} r\right) + D_{mn}\, K_n\!\left(\frac{m\pi}{\gamma} r\right)\right] \cos n\varphi \sin \frac{m\pi}{\gamma} z , \tag{33c}$$

where I_n and K_n are the modified BESSEL functions of the first and second kind, respectively.

Substitution of (33) into (30) yields expressions for the *static* displacements. The *static* stresses are then found from (16). The six constants of integration (four in the case $n = 0$) are found using the boundary conditions, (27) and (28).

C. Free vibration

We express the modal displacement vector, $\boldsymbol{U}$, in terms of a scalar potential, Φ, and a vector potential, $\boldsymbol{\Psi}$, i.e.,

$$\boldsymbol{U} = \nabla \Phi + \nabla \times \boldsymbol{\Psi} \tag{34}$$

where

$$\nabla \cdot \boldsymbol{\Psi} = 0\,, \qquad \boldsymbol{\Psi} = \Psi_r\, \boldsymbol{e}_r + \Psi_\varphi\, \boldsymbol{e}_\varphi + \Psi_z\, \boldsymbol{e}_z\,. \tag{35}$$

Substitution of (34) into (18) yields the following set of equations:

$$k^2 \nabla^2 \Phi + \Omega^2 \Phi = 0\,, \qquad \nabla^2 \boldsymbol{\Psi} + \Omega^2 \boldsymbol{\Psi} = 0\,. \tag{36}$$

According to DUHEM (see [27]), (34), (35), and (36) constitute a complete solution for $\boldsymbol{U}$.

We now assume the following form for the solutions of (36) which satisfy boundary conditions (27):

$$\Phi = \Phi^{mn}(r) \cos n\varphi \sin \frac{m\pi}{\gamma} z\,, \tag{37}$$

$$\Psi_r = \Psi_r^{mn}(r) \sin n\varphi \cos \frac{m\pi}{\gamma} z\,, \tag{38a}$$

$$\Psi_\varphi = \Psi_\varphi^{mn}(r) \cos n\varphi \cos \frac{m\pi}{\gamma} z\,, \tag{38b}$$

$$\Psi_z = \Psi_z^{mn}(r) \sin n\varphi \sin \frac{m\pi}{\gamma} z\,, \tag{38c}$$

and imposing (35) upon (38) we obtain

$$\frac{\Psi_r^{mn}}{r} + \frac{\partial \Psi_r^{mn}}{\partial r} - \frac{n}{r} \Psi_\varphi^{mn} + \frac{m\pi}{\gamma} \Psi_z^{mn} = 0\,. \tag{39}$$

Upon substitution of (37) and (38) into (36), and solving the resultant differential equations we obtain

$$\Phi^{mn} = A_1^{mn} Z_n(\alpha r) + B_1^{mn} W_n(\alpha r)\,, \tag{40}$$

$$\Psi_z^{mn} = A_2^{mn} Z_n(\beta r) + B_2^{mn} W_n(\beta r)\,, \tag{41a}$$

$$\Psi_r^{mn} - \Psi_\varphi^{mn} = 2\, A_3^{mn} Z_{n+1}(\beta r) + 2\, B_3^{mn} W_{n+1}(\beta r)\,, \tag{41b}$$

$$\Psi_r^{mn} + \Psi_\varphi^{mn} = 2\, A_4\, Z_{n-1}(\beta r) + 2\, B_4^{mn} W_{n-1}(\beta r)\,, \tag{41c}$$

where

$$\alpha^2 = \lambda_1 \left[\frac{\Omega^2}{k^2} - \left(\frac{m\pi}{\gamma}\right)^2\right] \quad \text{and} \quad \beta^2 = \lambda_2 \left[\Omega^2 - \left(\frac{m\pi}{\gamma}\right)^2\right]. \tag{42}$$

Z_n, W_n, λ_1, and λ_2 are defined by the following table:

Interval of Ω	λ_1	λ_2	$Z_n(\alpha r)$	$W_n(\alpha r)$	$Z_n(\beta r)$	$W_n(\beta r)$
$\frac{\Omega}{k} > \frac{m\pi}{\gamma}$	1	1	$J_n(\alpha r)$	$Y_n(\alpha r)$	$J_n(\beta r)$	$Y_n(\beta r)$
$\frac{\Omega}{k} < \frac{m\pi}{\gamma}$, $\Omega > \frac{m\pi}{\gamma}$	-1	1	$I_n(\alpha r)$	$K_n(\alpha r)$	$J_n(\beta r)$	$Y_n(\beta r)$
$\Omega < \frac{m\pi}{\gamma}$	-1	-1	$I_n(\alpha r)$	$K_n(\alpha r)$	$I_n(\beta r)$	$K_n(\beta r)$

J_n and Y_n are the BESSEL functions of the first and second kind, respectively, whereas I_n and K_n are the modified BESSEL functions of the first and second kind, respectively.

For the $n = 0$ (axisymmetric) terms (37) through (41) reduce to

$$\Phi = \Phi^{m0}(r) \sin \frac{m\pi}{\gamma} z\,, \qquad \Psi_\varphi = \Psi_\varphi^{m0}(r) \cos \frac{m\pi}{\gamma} z\,, \qquad \Psi_r = \Psi_z = 0 \tag{43}$$

where

$$\Phi^{m0} = A_5^{m0} Z_0(\alpha r) + B_5^{m0} W_0(\alpha r)\,, \qquad \Psi_\varphi^{m0} = A_6^{m0} Z_1(\beta r) + B_6^{m0} W_1(\beta r)\,. \tag{44}$$

Equation (39) is satisfied identically for $n = 0$.

The eigenfunctions, U_r^{mn}, U_φ^{mn}, and U_z^{mn}, are found by substituting (37), (38), (40), (41), (42), (43), and (44) into (34). For $n \geqq 1$, they are a function of eight constants of integration and for $n = 0$, they are a function of four constants. However, using (39) for $n \geqq 1$, the eigenfunctions can be made to be a function of only six constants.

The stress components due to modal displacements, $\tau_{rr}^{mn}, \tau_{\varphi\varphi}^{mn}, \tau_{zz}^{mn}, \tau_{rz}^{mn}, \tau_{r\varphi}^{mn}, \tau_{\varphi z}^{mn}$, are found by substituting the eigenfunctions into (21). Since stresses are specified at $r = a$ and $r = b$ for the forced motion problem, the following homogeneous stress boundary conditions must be satisfied by the modal stresses:

$$\begin{aligned} &\tau_{rr}^{mn}(a, \varphi, z) = \tau_{rz}^{mn}(a, \varphi, z) = \tau_{r\varphi}^{mn}(a, \varphi, z) = 0\,, \\ &\tau_{rr}^{mn}(b, \varphi, z) = \tau_{rz}^{mn}(b, \varphi, z) = \tau_{r\varphi}^{mn}(b, \varphi, z) = 0\,, \end{aligned} \quad n \geqq 1\,,$$
$$\begin{aligned} &\tau_{rr}^{m0}(a, z) = \tau_{rz}^{m0}(a, z) = 0\,, \\ &\tau_{rr}^{m0}(b, z) = \tau_{rz}^{m0}(b, z) = 0\,, \end{aligned} \quad n = 0\,. \tag{45}$$

Upon substitution of the modal stresses into (45), we obtain a homogeneous system of six linear algebraic equations in six unknowns for $n \geqq 1$ and a homogeneous system of four linear algebraic equations in four unknowns for $n = 0$.

(i) $n \geqq 1$

$$\begin{pmatrix} C_{ij} \\ i = 1, 2, 3, 4, 5, 6 \\ j = 1, 2, 3, 4, 5, 6 \end{pmatrix} \begin{Bmatrix} A_1^{mn} \\ B_1^{mn} \\ A_3^{mn} \\ B_3^{mn} \\ A_4^{mn} \\ B_4^{mn} \end{Bmatrix} = 0\,, \qquad C_{ij} = C_{ij}\left(\Omega_{mn}, \frac{h}{r}, \frac{L}{R}, \nu\right) \tag{46a}$$

(ii) $n = 0$

$$\begin{pmatrix} D_{ij} \\ i = 1, 2, 3, 4 \\ j = 1, 2, 3, 4 \end{pmatrix} \begin{pmatrix} A_5^{m0} \\ B_5^{m0} \\ A_6^{m0} \\ B_6^{m0} \end{pmatrix} = 0\,, \qquad D_{ij} = D_{ij}\left(\Omega_{m0}, \frac{h}{r}, \frac{L}{R}, \nu\right), \tag{46b}$$

A necessary and sufficient condition for the existence of a nontrivial solution for the modal coefficients A_1^{mn}, B_1^{mn}, etc., is that the determinants of the coefficient matrices in (46a) and (46b) vanish. For a given thickness ratio (h/R), length ratio (L/R) and POISSON's ratio, the resultant transcendental secular equations can be solved for the natural frequencies, Ω_{mnj}. The subscripts m and n ($m = 1, 2, 3, \ldots M - 1$ and $n = 1, 2, 3, \ldots N - 1$) denote the harmonic. For each harmonic (m, n) there will be an infinite set of frequencies denoted by the index $j = 1, 2, \ldots$.

In order to obtain a unique solution for the modal coefficients, A_1^{mnj}, B_1^{mnj}, A_3^{mnj}, B_3^{mnj}, A_4^{mnj}, B_4^{mnj} and A_5^{m0j}, B_5^{m0j}, A_6^{m0j}, B_6^{m0j}, we use the normalization condition, (22). Each eigenfunction and associated frequency is resubstituted into equations (46) and any one of them is deleted. The remaining five (three for $n = 0$) equations are then solved simultaneously with the normalization condition. The normalization condition in cylindrical coordinates for the problem being investigated is

$$1 = \int_a^b \int_0^\gamma \int_0^{2\pi} [(U_r^{mnj})^2 + (U_\varphi^{mnj})^2 + (U_z^{mnj})^2]\, r\, d\varphi\, dz\, dr\,. \tag{47}$$

Substitution of the eigenfunctions into (47) and subsequent integration, which can be completed in closed form, yields the additional equation needed for the determination of the modal coefficients.

We now have the expressions that give us the natural frequencies of vibrations and sufficient equations to uniquely determine the mode shape functions for a freely supported cylindrical shell.

D. Forced motion

With reference to (11), (23), and (26) we write the complete solution to the forced motion problem as

$$\boldsymbol{u}(r, \varphi, z, t) = \boldsymbol{u}^{(s)}(r, \varphi, z, t) + \sum_{m=1}^{M-1} \sum_{n=0}^{N-1} \sum_{j=1}^{\infty} \boldsymbol{U}^{mnj}(r, \varphi, z)\, q^{mnj}(t) \tag{48}$$

where

$$q^{mnj}(t) = q^{mnj}(0) \cos \Omega_{mnj} t + \frac{\dot{q}^{mnj}(0)}{\Omega_{mnj}} \sin \Omega_{mnj} t + \frac{1}{\Omega_{mnj}} \int_0^t Q^{mnj}(t_1) \sin \Omega_{mnj}(t - t_1)\, dt_1\,, \tag{49}$$

$$\Omega_{mnj}^2\, Q^{mnj}(t) = \int_S \left[\overset{n}{T}{}_r^{mnj} f_r + \overset{n}{T}{}_\varphi^{mnj} f_\varphi + \overset{n}{T}{}_\varphi^{mnj} f_z\right] ds - \int_S [U_r^{mnj} g_r + U_\varphi^{mnj} g_\varphi + U_z^{mnj} g_z]\, dS\,. \tag{50}$$

In view of (24), (25), and the initial conditions (29), we obtain

$$q^{mnj}(0) = Q^{mnj}(0+), \qquad \dot{q}^{mnj}(0) = \dot{Q}^{mnj}(0+). \tag{51a, b}$$

Upon examining the boundary conditions, (27) and (28), we see that

$$f_r = f_\varphi = f_z = g_\varphi = g_z = 0 \tag{52a}$$

everywhere and

$$g_r = - P_0 H(t) \frac{a_0 b_m}{2 \pi^2} \sin \frac{m \pi}{\gamma} z \qquad (n = 0), \tag{52b}$$

$$g_r = - P_0 H(t) \frac{a_n b_m}{\pi^2} \cos n \varphi \sin \frac{m \pi}{\gamma}, \qquad (n \geqq 1), \tag{52c}$$

at $r = b$ and equals zero everywhere else. Therefore,

(i) $n \geqq 1$

$$\Omega^2_{mnj} Q^{mnj}(t) = P_0 H(t) \frac{a_n b_m}{\pi^2} \int_0^\gamma \int_0^{2\pi} U_r^{mnj} \cos n \varphi \sin \left(\frac{m \pi}{\gamma} z \right) b \, d\varphi \, dz \tag{53a}$$

and

(ii) $n = 0$

$$\Omega^2_{m0j} Q^{m0j}(t) = P_0 H(t) \frac{a_0 b_m}{2 \pi^2} \int_0^\gamma \int_0^{2\pi} U_r^{m0j} \sin \left(\frac{m \pi}{\gamma} z \right) b \, d\varphi \, dz. \tag{53b}$$

Substitution of the eigenfunction, U_r^{mnj}, into (53) and integrating yields $Q^{mnj}(t)$. It is readily seen from (53) that

$$\ddot{Q}^{mnj}(t) = 0 \quad \text{for} \quad t > 0 \quad \text{and} \quad \dot{Q}^{mnj}(0+) = 0. \tag{54}$$

Hence, it follows from (49), (51), and (54) that

$$q^{mnj}(t) = Q^{mnj}(0+) \cos \Omega_{mnj} t. \tag{55}$$

$Q^{mnj}(0+)$ is obtained from (53) by replacing $H(t)$ by the numeral 1. Substitution of (55) into (48) along with the *static* solutions and the eigenfunctions yield the complete solution for the displacements. The stresses are obtained by using the stress-displacement relations, (7).

IV. A numerical example

A numerical example is now presented in order to obtain a basis for comparison between the elasticity solution and solutions obtained from the shell theories. A load of magnitude P_0 is suddenly applied radially inward over a small area on the outer surface of the shell. The dimensions of the loaded area are ε in the φ-direction and δ in the z-direction. The center of the loaded area is at the center of the lateral surface of the cylindrical shell ($\varphi = 0$, $z = \gamma/2$). The length-to-radius ratio (γ) is chosen to be 5.0 and the thickness-to-radius ratio is 0.1. This value of the thickness-to-radius ratio is chosen since most shell theories have been assumed to be valid up to this point in relatively long cylindrical shells. Poisson's ratio (ν) is chosen to be 0.3 since most manufacturing materials (aluminium and steel) fall near this value. The wave speed ratio, k, for this value of ν is $=1.871$.

The load distribution function for the example stated above is

$$f(\varphi) = \begin{cases} 0, & |\varphi| > \varepsilon/2 \\ 1, & |\varphi| \leqq \varepsilon/2 \end{cases}, \qquad f(\varphi + 2\pi) = f(\varphi),$$
$$g(z) = \begin{cases} 0, & 0 \leqq z < (\gamma - \delta)/2 \\ 1, & (\gamma - \delta)/2 \leqq z \leqq (\gamma + \delta)/2 \\ 0, & (\gamma + \delta)/2 < z \leqq \gamma. \end{cases} \tag{56}$$

The Fourier coefficients of these functions are

$$a_0 = \varepsilon, \qquad \bar{a}_n = \frac{2}{n} \sin \left(n \frac{\varepsilon}{2} \right), \qquad n = 1, 2, 3, \ldots,$$
$$\bar{b}_m = \frac{4}{m} \sin \frac{m \pi}{2} \sin \left(\frac{m \pi}{\gamma} \frac{\delta}{2} \right), \qquad m = 1, 3, 5, \ldots \tag{57}$$

We would like to choose ε, δ, M, and N so as to obtain a load geometry which would emphasize shear deformation and thickness stretch effects. Such a load geometry would be characterized by one where ε and δ are of the order of the thickness. However, as ε and δ are decreased, the number of terms (M, N) necessary to describe the load distribution (56) increases. After studying the effect several combinations of ε, δ, M and N

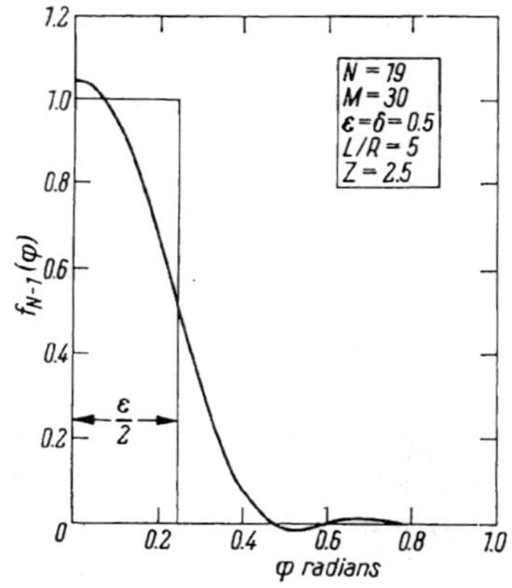

Fig. 2. Load distribution function vs. polar angle

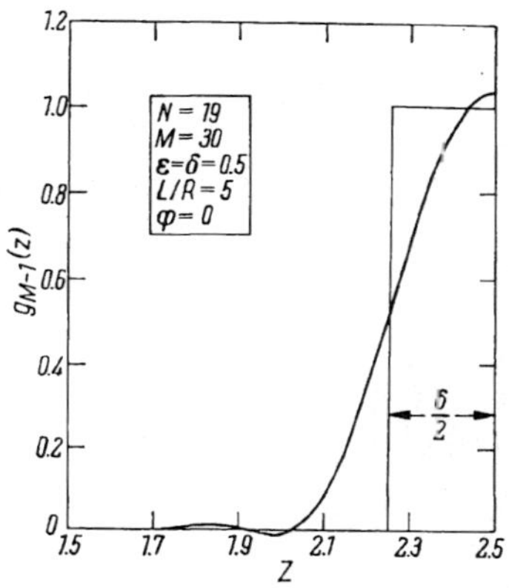

Fig. 3. Load distribution function vs. axial coordinate

had on the load distribution, the following numerical values of the parameters were chosen to keep the computation within limits.

$$\varepsilon = \delta = 0.5\,, \qquad N = 19\,, \qquad M = 30\,. \tag{58}$$

The resultant load distribution functions, $f_{N-1}(\varphi)$ and $g_{M-1}(z)$, are shown in Figures 2 and 3, respectively.

Reference [28] contains the solution to this numerical example for the FLÜGGE and improved shell theories.

The first step in determining the response of the shell is the calculation of the *static* solution. This is accomplished using the results of Section III B. The natural frequencies are then found, as indicated in Section III C, by setting the determinants of the coefficient matrices in (46) to zero. This yields transcendental equations for each two-fold harmonic (m, n), the roots of which are the desired frequencies. Following that, the modal coefficients corresponding to each natural frequency are determined by eliminating the first equation in each of (46a) and (46b) and solving the remaining equations of (46a) and (46b) with the normalization condition, (47). The eigenfunctions, which are now determined uniquely, are used to determine $q^{mnj}(t)$ as indicated by (53) and (55).

In the summations for the stresses and the displacements, there is an infinite sum for each two-fold harmonic on j. This apparent obstacle is overcome by calculating the hoop stress response on the external surface under the load ($\tau_{\varphi\varphi}$ vs. t for $r = 1.05$, $\varphi = 0$, $z = 2.5$) for the lowest 10 frequencies (summing $1 \leqq j \leqq 10$) for each two-fold harmonic. This calculation was then repeated using the lowest 15 frequencies (summing $1 \leqq j \leqq 15$) for each two-fold harmonic. The maximum of each of the calculations was then compared. This comparison showed that there was no difference at all in the first 6 significant figures. A stress, rather than a displacement, was used in the comparison, because stresses converge slower than the displacements. Because of this result, it was considered safe to truncate all the infinite series at $j = 15$.

Figures 4, 5, and 6 are representative of the forced motion response as calculated from the three theories. Figure 4 shows the radial displacement response under the load at the median surface. Figures 5 and 6 show the hoop and axial stress beneath the load on the external surface. The abscissa in the figures aforementioned is 40 dimensionless units of time, approximately one period of the radial displacement response. The solid line on each graph is the *static* solution as calculated from elasticity theory. The dynamic overshoot is approximately 40 to 50 percent. The elasticity theory and improved theory response agree to the order of one percent or less at their absolute maxima in all cases. The FLÜGGE theory response is slightly out of phase

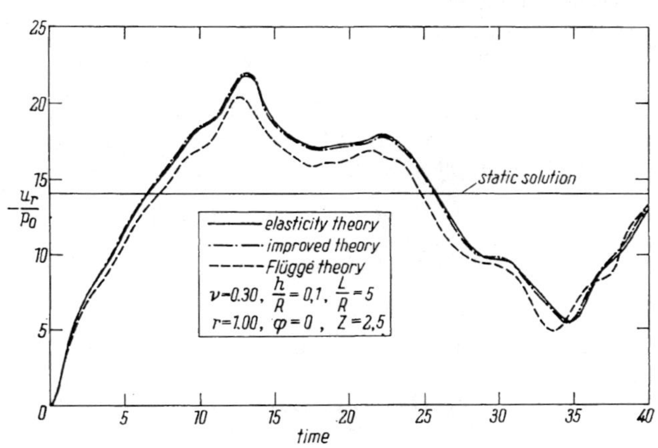

Fig. 4. Radial displacement vs. time

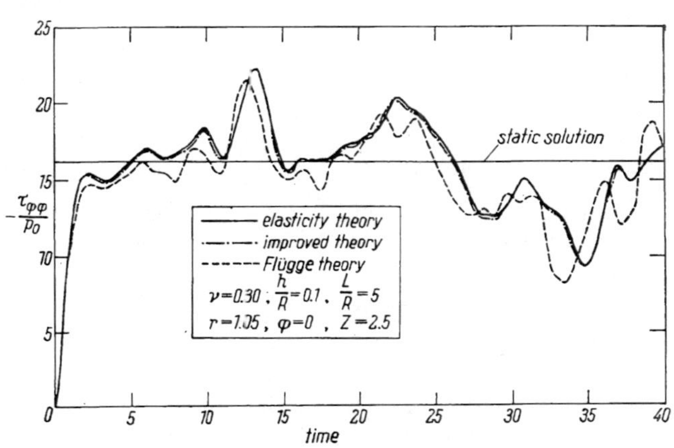

Fig. 5. Hoop stress vs. time

(faster response) and differs from the elasticity theory at the absolute maxima by approximately four to six percent.

Figure 7 shows the initial response of the hoop stress on the outer surface at the location $\varphi = \pi/2$, $z = 2.5$. This figure illustrates the wave nature of the systems of equations representing elasticity and improved theory. Elastic waves in solids travel with two distinct wave velocities according to linear elasticity theory.

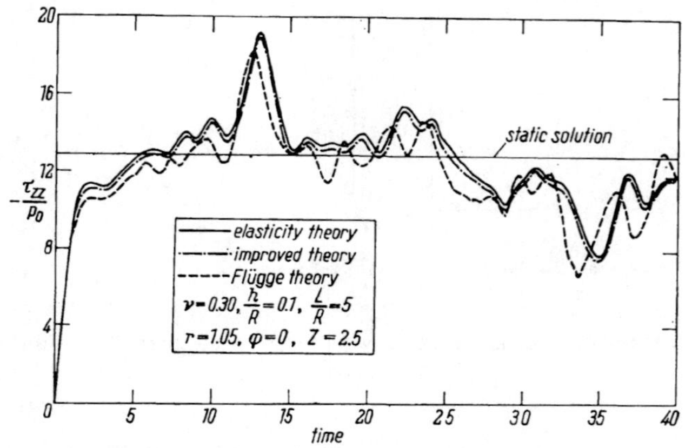

Fig. 6. Axial stress vs. time

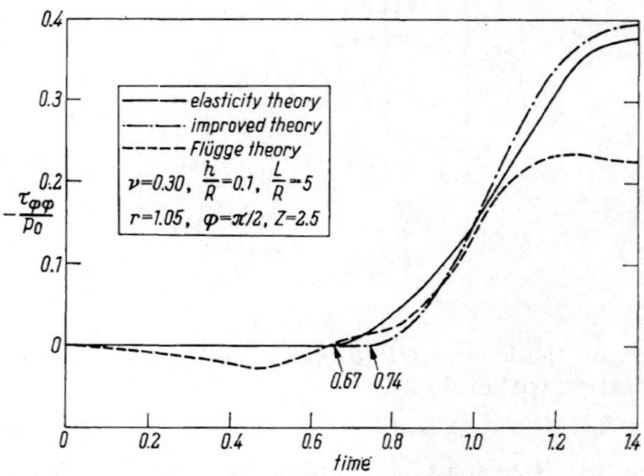

Fig. 7. Initial response: hoop stress vs. time

The faster one is called the dilational wave and travels with a velocity which equals $\sqrt{\frac{\lambda + 2\mu}{\varrho}} = \sqrt{\frac{E}{\varrho} \frac{(1-\nu)}{(1+\nu)(1-2\nu)}}$. The slower wave is called the shear wave and it travels with a velocity which equals $\sqrt{\frac{\mu}{\varrho}} = \sqrt{\frac{E}{2\varrho(1+\nu)}}$. Waves generated in improved theory solutions travel in the φ and z-directions with the slower membrane stress wave velocity which is equal to $\sqrt{\frac{E}{\varrho(1-\nu^2)}}$. There is immediate response through the thickness of the shell (See [21] and [22]). For this problem, using the non-dimensionalization in this paper, the elasticity theory predicts that the first wave should reach the point $r = 1.05$, $\varphi = \pi/2$, $z = 2.5$ at $\frac{(\pi/2) - (\varepsilon/2)}{k} \approx 0.71$. The improved theory response at this point should start at a time equal to $\left(\frac{\pi}{2} - \frac{\varepsilon}{2}\right) \Big/ \sqrt{\frac{2}{1-\nu}} \approx 0.78$. Upon examination of Figure 7, we see that the hoop stress determined by elasticity theory is zero until a time equal to 0.67, and when improved theory is used, it is zero until 0.74. The difference of 0.04 dimensionless units of time between the theoretical predictions and those of Figure 7 is probably due to the fact that the load does not have a sharp edge (see Figure 2). This figure also shows that solutions of Flügge's equation do not show a wave nature. This is due to the fact that Flügge theory possesses phase velocities which increase linearly with the harmonic (see references [21] and [22]). As a result of these extremely large phase velocities there is an immediate response.

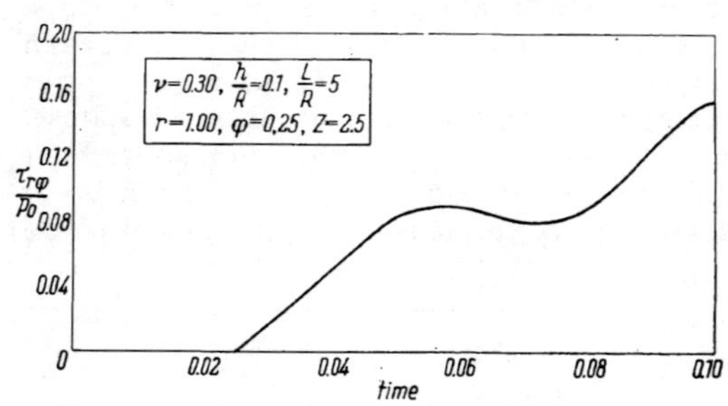

Fig. 8. Initial response: shear stress vs. time

Figure 8 is the early time history of the shear stress, $\tau_{r\varphi}$, at the edge of the load, perpendicular to the circumferential direction, at the median surface. According to elasticity theory, the time it takes for the dilational wave to go from the load to the median surface is $\frac{h}{2Rk} = .0267$. This is shown by the delayed response of the shear stresses in this figure. Neither improved or Flügge theory could have shown this time-delay because of the immediate response through the thickness that these theories predict.

V. Conclusions

The figures depicting the dynamic response for times containing the maximum displacements and stresses, show that the results predicted by improved and elasticity theory are within approximately one percent of each other. Their absolute maxima are roughly five percent greater than those determined by the Flügge theory. If accuracy is demanded, we should use improved theory for this period of time. Another reason for using the improved theory rather than elasticity theory in this time span is the computer time on the CDC 6400 necessary to obtain these results. The computing time for the FLÜGGE and improved theories was approximately two and four minutes, respectively, while for the elasticity theory it was approximately one hour.

Upon examination of the initial response results in Figures 7 and 8, we see the wave nature of elasticity and improved theory solutions. Due to the high phase velocities present in the FLÜGGE theory solution, there is an immediate response and consequently it results in a poor approximation in this time span.

In conclusion, we can say that the improved theory is an excellent approximation to the three dimensional theory of elasticity and gives suitable accuracy for the specific numerical case considered. We cannot generalize this conclusion to problems where high frequencies may be excited, such as a more localized load distribution or a load distribution which is an impulse in time. However, in all cases, care should be used when considering initial responses where the wave nature of an elastic material is important.

References

1 NAVIER, Memoire sur les lois de l'équilibre et du mouvement des corps solides élastiques, Paris, Memoires de l'Institut V. VII, pp. 375–393 (1827).
2 CHIU, Y. P., WEINSTEIN, A. S. and ZOROWSKI, C. F., The Stresses, Deformations and Bending of a Simply Supported Cylinder Subjected to a Symmetric Normal Surface Loading, Int. J. Mech. Sci. **5**, pp. 119–136 (1963).
3 VALOV, G. M., On the Axially-Symmetric Deformations of a Solid Circular Cylinder of Finite Length, PMM **26**, Nr. 4, pp. 650–667 (1962).
4 SUN, C. T. and VALANIS, K. C., On the Axially-Symmetric Deformations of a Hollow Cylinder of Finite Length Under the Action of Axially-Symmetric Loading, Developments in Mechanics **4**, pp. 171–185 (1967).
5 SIERAKOWSKI, R. L. and SUN, C. T., An Exact Solution to the Elastic Deformation of a Finite Length Hollow Cylinder, J. Franklin Inst., **286**, Nr. 2, pp. 99–113 (1968).
6 HERRMANN, G. and MIRSKY, I., Three-Dimensional and Shell-Theory Analysis of Axially Symmetric Motions of Cylinders, J. Appl. Mech., **23**, No. 4, pp. 563–568 (1956).
7 GAZIS, D. C., Three Dimensional Investigation of the Propagation of Waves in Hollow Circular Cylinders: Parts I and II, Acoust. Soc. Amer., **31**, Nr. 5, pp. 568–578 (1959).
8 GREENSPON, J. E., Flexural Vibrations of a Thick-Walled Circular Cylinder, Proce. Third U. S. Nat. Congr. Appl. Mech., pp. 163–173 June 1958.
9 GREENSPON, J. E., Flexural Vibrations of a Thick-Walled Circular Cylinder According to the Exact Theory of Elasticity, J. Aero/Space Sci. **27**, N. 1, pp. 37–40 (1960).
10 GREENSPON, J. E., Vibrations of a Thick-Walled Cylindrical Shell-Comparison of the Exact Theory with Approximate Theories, J. Acoust. Soc. Amer. **32**, Nr. 5, pp. 571–578 (1960).
11 GREENSPON, J. E., Axially Symmetric Vibrations of a Thick Cylindrical Shell in an Acoustic Medium, J. Acoust. Soc. Amer. **32**, Nr. 8, pp. 1017–1025 (1960).
12 GREENSPON, J. E., Vibrations of Thick and Thin Cylindrical Shells Surrounded by Water, J. Acoust. Soc. Amer. **33**, No. 10, pp. 1321–1328 (1961).
13 ARMENAKAS, A. E., On the Accuracy of Some Dynamic Shell Theories, J. Eng. Mech. Div. ASCE **93**, No. EM5, pp. 95–109 (1967).
14 ARMENAKAS, A. E., GAZIS, D. C. and HERRMANN, G., Free Vibrations of Circular Cylindrical Shells, First Edition, Pergamon Press, 1969.
15 MIRSKY, I., Wave Propagation in Transversely Isotropic Circular Cylinders, J. Acous. Soc. Amer., **37**, Nr. 6, pp. 1016–1026 (1965).
16 FLÜGGE, W., Statik und Dynamik der Schalen, Third Edition, Springer Verlag, Berlin, 1962.
17 VLASOV, V. Z., General Theory of Shells and Its Application in Engineering, NASA Technical Translation, NASA TT F-99, National Aeronautics and Space Administration, Washington, D. C., April 1964.
18 LOVE, A. E. H., A Treatise on the Mathematical Theory of Elasticity, Fourth Edition, Dover Publications, New York.
19 HERRMANN, G. and MIRSKY, I., Nonaxially Symmetric Motions of Cylindrical Shells, J. Acous. Soc. Amer., **29**, Nr. 10, pp. 1116–1123 (1957).
20 REISMANN, H. and MEDIGE, J., Dynamic Response of Cylindrical Shells (Part I), Report No. 13, Division of Interdisciplinary Studies and Research, School of Engineering, State University of New York at Buffalo.
21 PAWLIK, P. and REISMANN, H., Forced Plane Strain Motion of Cylindrical Shells — A Comparison of Shell Theory with Elasticity Theory, J. Appl. Mech., V. **40**, No. 3, pp. 125–130 (1973).
22 REISMANN, H. and PAWLIK, P., Plane-Strain Dynamic Response of a Cylindrical Shell — A Comparison Study of Three Different Shell Theories, J. Appl. Mech., **35**, N. 2, pp. 297–305 (1968).
23 LANCZOS, C., Discourse on Fourier Series, Hafner Publishing Company, New York, 1966, pp. 61–75.
24 REISMANN, H., On the Forced Motion of Elastic Solids, Appl. Scient. Res., **18**, pp. 156–165 (1967).
25 PAPCOVITCH, P. F., C. R., Acad. Sci., Paris, **195**, p. 513 (1932).
26 NEUBER, H., Kerbspannungslehre, J. Springer, Berlin, 1937.
27 STERNBERG, E., On the Integration of the Equations of Motion in the Classical Theory of Elasticity, Arch. Rational Mech. Anal., **6**, Nr. 1, pp. 34–50 (1960).
28 WEINGARTEN, L. I., Forced Motion of Cylindrical Shells — A Comparison of Shell Theory With Elasticity Theory, Ph. D. thesis, 1971, State University of New York at Buffalo, Buffalo.

Eingereicht am 24. 4. 1973

Anschriften: L. I. WEINGARTEN, Sandia Laboratories, Albuquerque, N. Mexico, USA
H. REISMANN, State University of New York at Buffalo, Buffalo, New York, USA

KLEINE MITTEILUNGEN

ZAMM 54, 193 (1974)

W. Gerlach

Nachkorrektur bei numerischer Differentiation*)

In [1] wurde ohne theoretische Begründung eine Faustregel mitgeteilt, die sich nach Meinung der Autoren gut zur Korrektur numerisch in Gleitkommaarithmetik berechneter Ableitungen eignet.

Da das Problem der Nachkorrektur numerisch berechneter Näherungen speziell bei der numerischen Differentiation von großer Wichtigkeit ist, soll hier eine Möglichkeit mitgeteilt werden, den Verfahrensfehler zu reduzieren. Der hier eingeschlagene Weg eignet sich allgemein zur Nachkorrektur bei parameterabhängigen Näherungsverfahren, so z. B. bei Quadraturverfahren, bei der Berechnung uneigentlicher Integrale und bei Iterationsverfahren.

Zur Berechnung der n-ten Ableitung einer Funktion $f(x)$ an einer fest vorgegebenen Stelle x_0 sei ein Näherungsverfahren $N_h(f)$ mit dem Diskretisierungsparameter h vorgelegt.

$F(x)$ sei eine Vergleichsfunktion, die an der Stelle x_0 wenigstens n-mal stetig differenzierbar ist und für die $F^{(n)}(x_0)$ einfach berechenbar ist. Wir betrachten die Größe $N_{\bar{h}}^1(f)$, welche definiert ist durch

$$N_{\bar{h}}^1(f) = N_h(f) + \frac{F^{(n)}(x_0) - N_h(F)}{N_h(F) - N_{\bar{h}}(F)} (N_h(f) - N_{\bar{h}}(f)) . \tag{1}$$

Dabei sei $\bar{h} = \bar{h}(h)$, und es gelte

$$\lim_{h\to 0} \bar{h}(h) = 0 , \quad \bar{h} \neq h \quad \text{und} \quad N_h(F) \neq N_{\bar{h}}(F) .$$

Lemma: *Falls die Beziehung*

$$\lim_{h\to 0} \frac{f^{(n)}(x_0) - N_{\bar{h}}(f)}{f^{(n)}(x_0) - N_h(f)} = \lim_{h\to 0} \frac{F^{(n)}(x_0) - N_{\bar{h}}(F)}{F^{(n)}(x_0) - N_{\bar{h}}(F)} = q \neq 1 \tag{2}$$

erfüllt ist, gilt

$$\lim_{h\to 0} \frac{f^{(n)}(x_0) - N_{\bar{h}}^1(f)}{f^{(n)}(x_0) - N_h(f)} = 0 . \tag{3}$$

Der Beweis kann auf direktem Wege unter Benutzung von (1) und (2) unmittelbar erbracht werden.

Die Beziehung (2) ist ohne wesentliche Voraussetzungen über F z. B. stets erfüllt, wenn N_h ein Verfahren der Ordnung k ($k > 0$) ist, d. h. wenn für jede hinreichend oft stetig differenzierbare Funktion $g(x)$ gilt

$$g^{(n)}(x_0) - N_h(g) = c\, h^k\, g^{(n+k)}(\xi) \tag{4}$$

und außerdem zwischen h und $\bar{h}$ eine Beziehung der Form $\bar{h} = a\, h$ ($a \neq 0, 1$) besteht.

Ist die Beziehung (2) erfüllt, dann stellt $N_{\bar{h}}^1(f)$ für genügend kleines h eine bessere Näherung für $f^{(n)}(x_0)$ dar als $N_h(f)$ und zwar ohne Kenntnis der Verfahrensordnung k, wie sie z. B. bei Anwendung des ebenfalls in [1] betrachteten Richardson-Romberg-Prinzips notwendig ist.

Das Richardson-Romberg-Prinzip ist in (1) als Spezialfall enthalten. Setzt man bei Kenntnis der Verfahrensordnung $F = x^{n+k}$, so ergibt sich wegen

$$\frac{F^{(n)}(x_0) - N_h(F)}{N_h(F) - N_{\bar{h}}(F)} = \frac{-1}{1 - [F^{(n)}(x_0) - N_{\bar{h}}(F)]/[F^{(n)}(x_0) - N_h(F)]}$$

$$= \frac{-1}{1 - [(n+k)!\, \bar{h}^k]/[(n+k)!\, h^k]} = - \frac{1}{1 - (\bar{h}/h)^k}$$

mit (1) die bekannte Korrekturvorschrift des Richardson-Romberg-Prinzips

$$N_{\bar{h}}^1(f) = N_h(f) - \frac{1}{1 - (\bar{h}/h)^k} (N_h(f) - N_{\bar{h}}(f)) . \tag{5}$$

*) Die Veröffentlichung geschieht mit Zustimmung der Verfasser der unter [1] zitierten Arbeit.

Bei der Anwendung der Beziehung (1) ist zu beachten, daß bei zu klein gewähltem h die Rundungsfehler dominieren und das Ergebnis völlig verfälschen können. Aus diesem Grunde ist für praktische Rechnungen die Aussage (3) abzuändern in

$$\left|\frac{f^{(n)}(x_0) - N_{\bar{h}}^1(f)}{f^{(n)}(x_0) - N_h(f)}\right| < 1 \qquad (0 < \varepsilon_1 < |h| < \varepsilon_2) . \tag{3.1}$$

Beispiele:

1. $f(x) = \cot x$, $\quad x_0 = 0.04$

$$N_h(f) = \frac{1}{2\,h} (f(x_0 + h) - f(x_0 - h)) \quad (k = 2,\ n = 1)$$

(Gleitkommarechnung mit 9 Mantissenstellen)

a) $h = 0{,}01$, $\bar{h} = 0{,}02$

$N_h(f)$	$N_{\bar{h}}(f)$	$N_{\bar{h}}^1(f)$	F
−667,000115	−833,666790	−611,444557	x^3
		−620,112215	$\ln x$
		−625,333446	$1/x$

b) $h = 0{,}005$, $\bar{h} = 0{,}01$

$N_h(f)$	$N_{\bar{h}}(f)$	$N_{\bar{h}}^1(f)$	F
−635,254067	−667,000115	−624,672051	x^3
		−625,072519	$\ln x$
		−625,333427	$1/x$

exakter Wert: −625,333447

2. $f(x) = \exp(x)$, $\quad x_0 = 1$,

$$N_h(f) = \frac{1}{h} (f(x_0 + h) - f(x_0)) \qquad (k = 1,\ n = 1)$$

(Festkommarechnung mit 6 Stellen nach dem Komma)

a) $h = 0{,}01$, $\bar{h} = 0{,}02$

$N_h(f)$	$N_{\bar{h}}(f)$	$N_{\bar{h}}^1(f)$	F
2,731900	2,745650	2,718150	x^2
		2,718222	$\sinh x$
		2,718417	$\frac{1}{2-x}$

b) $h = 0{,}005$, $\bar{h} = 0{,}02$

$N_h(f)$	$N_{\bar{h}}(f)$	$N_{\bar{h}}^1(f)$	F
2,725000	2,745650	2,723623	x^2
		2,718265	$\sinh x$
		2,718295	$\frac{1}{2-x}$

exakter Wert: 2,718282

Literatur

1 Filippi, S. und Engels, H., Altes und Neues zur numerischen Differentiation, Elektron. Datenverarbeitung (1966), S. 57–65.

Eingereicht am 2. 10. 1972 (rev. Fassung: 24. 5. 1973)

Anschrift: Dr. W. Gerlach, Bergakademie Freiberg, Sektion Mathematik, 92 Freiberg, Bernhard-von-Cotta-Str. 2, DDR

ZAMM 54, 194 – 196 (1974)

B. D. Aggarwala

On a Monte Carlo Method for Rectilinear Plates

1. Introduction

There have been some attempts lately to devise Monte Carlo methods for the solution of plate problems in elasticity [3], [4], [5]. The purpose of this note is to outline a Monte Carlo method for a simply supported rectilinear plate lying on an elastic foundation.

It is to be noted that quite often, in a plate problem, one is not really interested in the values, or behaviour, of the solution all through the plate and deflection, or moments, or both, at some specified points are the only information needed. The method outlined below gives the values of w and Δw (see below for definition of these quantities) at desired points directly. Similar methods have been used in heat flow problems by Haji-Sheikh and Sparrow [6] and in plate problems by Gopalsamy and Aggarwala [4].

2. The problem

We wish to solve the following boundary value problem for a plane region D, with boundary ∂D.

$$\Delta\Delta w + \alpha\,\Delta w + \beta\, w = 0 \qquad \text{in} \quad D \tag{1a}$$

with

$$w = f_1(s) \qquad \text{on} \quad \partial D \tag{1b}$$

and

$$\Delta w = f_2(s) \qquad \text{on} \quad \partial D\,. \tag{1c}$$

Here w is the unknown function to be determined, Δ is the two-dimensional Laplacian operator ($\Delta = \partial^2/\partial x^2 + \partial^2/\partial y^2$), α and β are real constants and $f_1(s)$ and $f_2(s)$ are specified functions on ∂D.

3. A mean value relation

We write equation (1a) in the form

$$(\Delta - k_1^2)\,(\Delta - k_2^2)\, w = 0 \tag{2}$$

where k_1 and k_2 are suitable constants.

Now consider a circle of radius r with center at $P_0(x_0, y_0)$ contained in D. In this circle, the solution of equation (2) may be written in the form

$$w = A_0 I_0(k_1 r) + B_0\, I_0(k_2 r) + \sum_{n=1}^{\infty} [A_n I_n(k_1 r) + B_n I_n(k_2 r)] \times$$

$$\times \sin n\,\theta + \sum_{n=1}^{\infty} [C_n\, I_n(k_1 r) + D_n\, I_n(k_2 r)] \cos n\,\theta \tag{3}$$

where I_n $(n \geqq 0)$ is the modified Bessel function of order n. Equation (3) gives

$$\frac{1}{2\pi}\int_0^{2\pi} w(P)\, d\theta = A_0\, I_0(k_1 r) + B_0\, I_0(k_2 r) \tag{4}$$

and

$$\frac{1}{2\pi}\int_0^{2\pi} \Delta w(P)\, d\theta = k_1^2\, A_0\, I_0(k_1 r) + k_2^2\, B_0\, I_0(k_2 r) \tag{5}$$

where P is a point on the circle. Also, we have

$$w_0 = A_0 + B_0\,, \tag{6}$$

$$\Delta w_0 = k_1^2\, A_0 + k_2^2\, B_0 \tag{7}$$

where w_0 and Δw_0 are the values of w and Δw, respectively, at the center of the circle. From equations (4), (5), (6) and (7) we obtain

$$w_0 = \frac{1}{2\pi}\, F_1(r) \int_0^{2\pi} w(P)\, d\theta + \frac{1}{2\pi}\, F_2(r) \int_0^{2\pi} \Delta w(P)\, d\theta \tag{8a}$$

and

$$\Delta w_0 = \frac{1}{2\pi}\, F_3(r) \int_0^{2\pi} w(P)\, d\theta + \frac{1}{2\pi}\, F_4(r) \int_0^{2\pi} \Delta w(P)\, d\theta \tag{8b}$$

where

$$F_1(r) = \frac{k_1^2\, I_0(k_1 r) - k_2^2\, I_0(k_2 r)}{(k_1^2 - k_2^2)\, I_0(k_1 r)\, I_0(k_2 r)}\,, \tag{9a}$$

$$F_2(r) = -\frac{I_0(k_1 r) - I_0(k_2 r)}{(k_1^2 - k_2^2)\, I_0(k_1 r)\, I_0(k_2 r)}\,, \tag{9b}$$

$$F_3(r) = \frac{k_1^2\, k_2^2\, (I_0(k_1 r) - I_0(k_2 r))}{(k_1^2 - k_2^2)\, I_0(k_1 r)\, I_0(k_2 r)}\,, \tag{9c}$$

and

$$F_4(r) = -\frac{k_2^2\, I_0(k_1 r) - k_1^2\, I_0(k_2 r)}{(k_1^2 - k_2^2)\, I_0(k_1 r)\, I_0(k_2 r)}\,. \tag{9d}$$

We conclude that the solutions w of equation (2) satisfy the mean value relations (8).

Conversely, (see also section 6 below) let v and w be two functions which satisfy the following mean value relations at all voints $P_0 \in D$.

$$w(P_0) = \frac{1}{2\pi}\, F_1(r) \int_0^{2\pi} w(P)\, d\theta + \frac{1}{2\pi}\, F_2(r) \int_0^{2\pi} v(P)\, d\theta\,, \tag{10a}$$

$$v(P_0) = \frac{1}{2\pi}\, F_3(r) \int_0^{2\pi} w(P)\, d\theta + \frac{1}{2\pi}\, F_4(r) \int_0^{2\pi} v(P)\, d\theta\,. \tag{10b}$$

We assume that these relations are satisfied by v and w for all values of r such that the resulting circle is contained in D. We further assume that v and w are continuously differentiable a sufficient number of times. If now, we substitute the following relation [2], p. 288 in equations (10)

$$\frac{1}{2\pi}\int_0^{2\pi} v(P)\, d\theta = v(P_0) + \frac{1}{4}\, r^2 (\Delta v)_{P_0} + O(r^3) \tag{11}$$

and a similar relation for w and equate the powers of r on both sides, the following relations are obtained

$$v(P_0) = (\Delta w)_{P_0}\,, \tag{12a}$$

$$(\Delta v)_{P_0} - (k_1^2 + k_2^2)\, v(P_0) + k_1^2\, k_2^2\, w(P_0) = 0\,. \tag{12b}$$

Since P_0 is an arbitrary point in D, equations (12) are satisfied everywhere in D and thus are seen to be equivalent to equation (2). This proves the equivalence of equations (2) and (10) under appropriate conditions.

4. The Monte Carlo Procedure

Based on equations (10), we now devise a Monte Carlo procedure for the solution of our problem.

Let $P_0(x_0, y_0)$ be the point at which the solution is desired. A random walking particle starts from (x_0, y_0) and chooses its subsequent positions in the following manner: From (x_0, y_0) as center, we construct a circle of radius R_1, where R_1 is the maximum distance from (x_0, y_0) to the boundary of D. The particle now assumes a position on the boundary of this circle chosen in a uniformly random manner, each point on the boundary of the circle being equally likely. Let (x_1, y_1) be the new position. The process is repeated with (x_1, y_1) as the center of a new circle. When the particle "reaches the boundary", it is absorbed and the walk is terminated, (see ref. [6] and also section (5) below).

We now associate two random variables, $z_1(x_k, y_k)$ and $z_2(x_k, y_k)$ with this particle. The values of these variables for any given walk are defined in the following manner: Let (x_k, y_k) and (x_{k+1}, y_{k+1}) be two successive positions of the particle. Then we define

$$z_1(x_k, y_k) = F_1(R_{k+1})\, z_1(x_{k+1}, y_{k+1}) +$$

$$+ F_2(R_{k+1})\, z_2(x_{k+1}, y_{k+1})\,, \tag{13a}$$

$$z_2(x_k, y_k) = F_3(R_{k+1})\, z_1(x_{k+1}, y_{k+1}) + F_4(R_{k+1})\, z_2(x_{k+1}, y_{k+1}) \tag{13b}$$

where F_1, F_2, F_3 and F_4 are defined in equations (9). If the walk reaches the point $P_n(x_n, y_n)$ on the boundary, we define

$$z_1(x_n, y_n) = f_1(P_n)\,, \qquad z_2(x_n, y_n) = f_2(P_n)\,. \tag{14}$$

This completes the definition of $z_1(x_k, y_k)$ and $z_2(x_k, y_k)$.

Now let the particle jump from the point (x_k, y_k) to the point (x_{k+1}, y_{k+1}). Under this condition, equations (13) give

$$E[z_1(x_k, y_k)] = F_1(R_{k+1})\, E[z_1(x_{k+1}, y_{k+1})] + \\ + F_2(R_{k+1})\, E[z_2(x_{k+1}, y_{k+1})]\,, \tag{15a}$$

$$E[z_2(x_k, y_k)] = F_3(R_{k+1})\, E[z_1(x_{k+1}, y_{k+1})] + \\ + F_4(R_{k+1})\, E[z_2(x_{k+1}, y_{k+1})] \tag{15b}$$

where $E[z]$ denotes the expectation of z. It should be noted that if (x_k, y_k) is known, then R_{k+1} is also known.

Equations (15) give the conditional expectation of $z_1(x_k, y_k)$ and $z_2(x_k, y_k)$, subject to the condition that the next position is (x_{k+1}, y_{k+1}). However, since every position on the circle of radius R_{k+1} is equally likely, the unconditional expectations of $z_1(x_k, y_k)$ and $z_2(x_k, y_k)$ are given by

$$E[z_1(x_k, y_k)] \\ = \frac{1}{2\pi} F_1(R_{k+1}) \int_0^{2\pi} E[z_1(x_k + R_{k+1}\cos\theta, y_k + R_{k+1}\sin\theta)]\, d\theta + \\ + \frac{1}{2\pi} F_2(R_{k+1}) \int_0^{2\pi} E[z_2(x_k + R_{k+1}\cos\theta, y_k + R_{k+1}\sin\theta)]\, d\theta\,, \tag{16a}$$

$$E\,[z_2(x_k, y_k)] \\ = \frac{1}{2\pi} F_3(R_{k+1}) \int_0^{2\pi} E\,[z_1\,(x_k + R_{k+1}\cos\theta, y_k + R_{k+1}\sin\theta)]\, d\theta + \\ + \frac{1}{2\pi} F_4(R_{k+1}) \int_0^{2\pi} E\,[z_2\,(x_k + R_{k+1}\cos\theta, y_k + R_{k+1}\sin\theta)]\, d\theta\,. \tag{16b}$$

Equations (16) are similar to equations (10), (see also section 6 below). We shall now consider the boundary conditions: We note that, for small r, $F_1 = 1 + O(r^2)$, $F_2 = O(r^2)$, $F_3 = O(r^2)$ and $F_4 = 1 + O(r^2)$, so that as $r \to 0$, the relations (13) tend to become

$$z_1(x_k, y_k) = z_1(x_{k+1}, y_{k+1}),\; z_2(x_k, y_k) = z_2(x_{k+1}, y_{k+1}) \tag{17}$$

which imply that if all R_k are small in a walk, then $z_1(P_0) \simeq f_1(P_n)$ and $z_2(P_0) \simeq f_2(P_n)$ where P_0 is the starting and P_n is the finishing point of the walk. Also, if the starting point P_0 of the random walking particle is close to the boundary, then R_1 and all the subsequent values of R are likely to be small, since the probability that a particle which starts close to the boundary will jump into the interior of D is small. Also, if all R's in a walk are small, then P_n is likely to be close to P_0. These arguments show that as P_0 approaches a point Q on the boundary, $E[z_1(P_0)]$ and $E[z_2(P_0)]$ approach $f_1(Q)$ and $f_2(Q)$ respectively.

To clarify this point further, let us consider the case of the problem [6]:

$$\Delta u = 0 \quad \text{in } D \qquad \text{with} \qquad u = f(s) \text{ on } \partial D.$$

In this case, the floating random walk follows exactly the same procedure as above. The variables z_1 and z_2 are replaced by a single variable (call it z) and the equations (13) are replaced by the single equation $z(x_k, y_k) = z(x_{k+1}, y_{k+1})$ so that if P_0 and P_n are the starting and the finishing point of a walk respectively, and if P_0 is close to the boundary ∂D, one would expect P_0 to be close to P_n for the boundary condition to be satisfied and this in turn would necessitate that if P_0 is close to the boundary ∂D, then all R's in the random walk are small. Since the validity of a floating random walk for a harmonic problem has been well established, we conclude that the above arguments in our case are valid.

It remains to calculate $E[z_1(P_0)]$ and $E[z_2(P_0)]$. Sample values of $z_1(P_0)$ and $z_2(P_0)$ can be easily calculated with the help of equations (13) and (14). If P_n is the point reached on the boundary, equations (14) give $z_1(P_n)$ and $z_2(P_n)$ and we use equations (13) to calculate $z_1(P_{n-1})$, $z_2(P_{n-1})$, $z_1(P_{n-2})$, $z_2(P_{n-2})$, ..., $z_1(P_0)$, $z_2(P_0)$ successively. Sufficiently large number of such random walks give close estimates of $E[z_1(P_0)]$ and $E[z_2(P_0)]$.

5. Numerical results

In the numerical results reported, we calculate the central deflection and sum of bending moments at the center of a uniformly loaded, simply supported, rhombic plate lying on an elastic foundation. For this problem, the relevant equation is $\Delta\Delta w + \beta\, w = 1$ with $w = \Delta w = 0$ at the boundary, and we report the values of w_0 and Δw_0 at the center of the plate for various values of β and various values of φ, half the acute angle of the rhombus. The smaller of the two diagonals of the rhombus is taken to be of length two. A strip of width .01 which is .005 units wide on each side of the boundary is constructed around ∂D to facilitate landing of the random walker on ∂D. If the walker lands in the strip, it is assumed to have landed on ∂D. The number of walks in each case is 5000 and the means μ_1 and μ_2 of z_1 and z_2 respectively from these walks are reported in tables 1 and 2. Let s_1 and s_2 be the standard deviations of $z_1(P_0)$ and $z_2(P_0)$ respectively, i.e., $s_i = \sqrt{\sum_{j=1}^{5000} (\mu_i - z_{ij})^2/4999}$, where z_{ij}, $j = 1, \ldots, 5000$ are the sample values of z_i, $i = 1, 2$. The quantities μ_1 and μ_2 are distributed around w_0 and Δw_0 with standard deviations $\sigma_1 = s_1/\sqrt{5000}$ and $\sigma_2 = s_2/\sqrt{5000}$ respectively. The quantities σ_1 and σ_2, thus

Table 1
Values of $\mu_1 = E\,[10^4\, z_1]$ and its standard deviation σ_1, at the center of a rhombus

φ	β: 1	10	50	100	500	1000	10,000
45°	$\mu_1 =$ 162.3	148.6	107.4	78.21	23.91	12.08	.9928
	$\sigma_1 =$ 1.88	1.51	.649	.294	.026	.012	.0002
40°	$\mu_1 =$ 224.6	199.2	130.5	90.18	24.26	11.83	.9885
	$\sigma_1 =$ 2.68	1.91	.654	.269	.025	.012	.0001
35°	$\mu_1 =$ 298.7	257.4	153.5	100.1	24.16	11.46	.9897
	$\sigma_1 =$ 3.48	2.35	.684	.256	.025	.012	.0001
30°	$\mu_1 =$ 401.4	327.4	173.9	107.5	23.70	11.14	.9918
	$\sigma_1 =$ 5.51	3.02	.693	.241	.027	.012	.0001
25°	$\mu_1 =$ 531.9	401.4	191.3	112.1	23.12	10.86	.9940
	$\sigma_1 =$ 7.46	3.79	.701	.229	.030	.01	.0001
20°	$\mu_1 =$ 694.6	478.2	203.4	114.4	22.64	10.67	.9957
	$\sigma_1 =$ 10.95	4.41	.684	.213	.030	.01	.0001
15°	$\mu_1 =$ 857.7	568.2	212.8	115.2	22.20	10.51	.9969
	$\sigma_1 =$ 12.97	5.14	.671	.214	.030	.009	.0001

Table 2
Values of $\mu_2 = E\,[10^4\, z_2]$ and its standard deviation σ_2, at the center of a rhombus

φ	β: 1	10	50	100	500	1000	10,000
45°	$\mu_2 =$ 1465	1328	920.5	650.9	134.7	35.56	−2.659
	$\sigma_2 =$ 6.75	4.47	1.99	1.99	1.02	.509	0.30
40°	$\mu_2 =$ 1718	1505	936.1	613.6	91.42	13.97	−1.408
	$\sigma_2 =$ 7.96	4.55	2.43	2.36	.924	.395	.022
35°	$\mu_2 =$ 1980	1667	921.0	551.5	56.27	−.2197	−.5903
	$\sigma_2 =$ 9.14	4.64	2.93	2.84	.837	.3090	.0138
30°	$\mu_2 =$ 2273	1811	870.7	484.0	31.60	−7.343	−.1926
	$\sigma_2 =$ 11.61	4.82	3.96	3.14	.741	.233	.0079
25°	$\mu_2 =$ 2597	1907	806.3	408.9	15.13	−9.685	.0151
	$\sigma_2 =$ 14.4	5.30	4.59	3.50	.639	.181	.0046
20°	$\mu_2 =$ 2927	1982	718.7	341.2	6.798	−10.52	.1093
	$\sigma_2 =$ 17.45	5.81	5.48	3.73	.504	.164	.0033
15°	$\mu_2 =$ 3239	2025	626.7	275.3	.3707	−10.50	.1441
	$\sigma_2 =$ 19.8	0.66	6.12	3.84	.4253	.160	.0028

calculated, are also reported in the tables. It should be noted that if we consider that $|w_0 - E[z_1(P_0)]|$ and $|\Delta w_0 - E[z_2(P_0)]|$ are likely to be less than $2\sigma_1$ and $2\sigma_2$ respectively, this would give an accuracy of better than 2% in most cases reported. The values reported are compared with those available in the literature. The values for a square, for various values of β, compare well with series results [8] and the error in all cases is less than 2%, (table 3). Our results for $\beta = 1$ compare well with MORLEY's results [7] for $\beta = 0$ (table 4). The slightly larger deviations are undoubtedly due to the difference in the corresponding values of β. Numerical results seem to indicate that our method does not give good results for $\beta \ll 1.0$ (see also [4]).

Table 3
Comparison of author's results with series results [8] for a square

$\beta =$		1	10	50	100	500	1000	10,000
$10^4 w_0$	Author	162.3	148.6	107.4	78.21	23.91	12.08	.9928
	Series	160.8	147.0	106.1	78.36	23.93	12.09	.9931
$10^4 \Delta w_0$	Author	1465	1328	920.5	650.9	134.7	35.56	−2.659
	Series	1457	1321	920.4	649.4	133.6	35.90	−2.681

Table 4
Comparison of author's results for $\beta = 1$ with MORLEY's results for $\beta = 0$, [7]

$\varphi =$		40°	30°	25°	20°	15°
$10^4 w_0$	Author	224.6	401.4	531.9	694.6	857.7
	MORLEY	226.7	409.6	539.2	700.1	909.2
$10^4 \Delta w_0$	Author	1718	2273	2597	2927	3239
	MORLEY	1739	2332	2671	3031	3433

It is perhaps worth pointing out that for large values of β, the quantity Δw becomes negative near the center. This phenomenon is well known in problems in channel flow [1]. As β further increases, the plate becomes almost flat near the center where $w_0 \approx 1/\beta$ and $\Delta^2 w_0 \approx 0$.

6. Further discussion on Mean Value Theorem

It should be noted that the converse of Mean Value Theorem proved in section 3 above does not properly justify the Monte Carlo Method. While proving the converse of Mean Value Theorem we assumed that equations (10) are satisfied for *all* sufficiently small r at every point $P_0 \in D$ while the Monte Carlo Method assumes that the M. V. T. is satisfied for only one radius $r(P_0)$ at every point $P_0 \in D$. For the particular case of $\operatorname{Re} k_1^2 \geqq 0$, $\operatorname{Re} k_2^2 \geqq 0$ in equation (2), (where $\operatorname{Re} z$ denotes the real part of z), we prove the following theorem which assumes that M. V. T. is satisfied for only one radius $r(P_0)$ at every point $P_0 \in D$. First we prove the following lemma.

Lemma: *If* $|\operatorname{Arg} k| \leqq \pi/4$, *then* $|I_0(k\,r)| \geqq 1$ *for all* $r \geqq 0$.

Proof: It is known that [9], p. 82, $|I_0(z)| \geqq 1$ if $|\operatorname{Arg} z| = \pi/4$, and that [9], p. 203, $|I_0(z)| \approx |e^z/\sqrt{2\pi z}|$ for large $|z|$, if $\operatorname{Re} z > 0$. Now apply the Principle of Maximum Modulus to $1/I_0(z)$ where z lies in the sector bounded by a sufficiently large circular arc $|z| = r$, and the lines $\operatorname{Arg} z = \pm \pi/4$.

Theorem: Hypotheses

1. *D is a bounded region in the plane with boundary ∂D and such that the boundary value problem* (1) *is solvable in* $D \cup \partial D$.
2. *v_1 and w_1 are real and continuous in* $D \cup \partial D$.
3. $\operatorname{Re} k_1^2 \geqq 0$ *and* $\operatorname{Re} k_2^2 \geqq 0$ *in equation* (2).
4. *Corresponding to every point $P_0 \in D$, there is defined a number $r(P_0) > 0$, such that v_1 and w_1 satisfy the mean value relations* (10) *for $r = r(P_0)$ at the point P_0, and such that the corresponding circle is contained in* $D \cup \partial D$.

Conclusion:

v_1 and w_1 satisfy equations (12) *for all* $P_0 \in D$.

Proof: Consider the function w_2 which satisfies equation (1a) with $w_2 = w_1$ and $\nabla^2 w_2 = v_1$ on ∂D. It is clear that the functions $v_3 = \nabla^2 w_2 - v_1$ and $w_3 = w_2 - w_1$ vanish on ∂D and satisfy the mean value relations (10) for $r = r(P_0)$. Now consider the functions $v = v_3 - k_2^2 w_3$ and $w = w_3$. From the fact that v_3 and w_3 satisfy equations (10), it is easy to obtain the mean value relations for v and w which hold for $r = r(P_0)$. These turn out to be

$$v_0 = \frac{1}{2\pi I_0(k_1 r)} \int_0^{2\pi} v(P)\, d\theta\,, \tag{18a}$$

$$w_0 = \frac{1}{2\pi I_0(k_2 r)} \int_0^{2\pi} w(P)\, d\theta + \frac{F_2(r)}{2\pi} \int_0^{2\pi} v(P)\, d\theta\,. \tag{18b}$$

We shall prove that $v = w = 0$ for all $P \in D$ which will prove the theorem. If possible let $v \neq 0$ somewhere in D. Since v and hence $|v|$ is continuous in $D \cup \partial D$, the set S of points where $|v|$ assumes its maximum in $D \cup \partial D$ is closed. Let P_0 be a point in S which is closest to ∂D. Let $P_0 \in D$. Now for $r = r(P_0) > 0$, we have

$$\frac{1}{2\pi} \int_0^{2\pi} v(P)\, d\theta = I_0(k_1 r)\, v(P_0)$$

which implies that

$$\frac{1}{2\pi} \int_0^{2\pi} |v(P)|\, d\theta \geqq |I_0(k_1 r)|\; |v(P_0)|$$

and since $|I_0(k_1 r)| \geqq 1$, this implies a contradiction. It follows that $|v|$ assumes its maximum on ∂D. But $|v| = 0$ on ∂D. Therefore, $v = 0$ in D. The mean value relation (18b) now becomes $(1/2\pi) \int_0^{2\pi} w(P)\, d\theta = I_0(k_2 r)\, w(P_0)$ and the same argument applied to w shows that $w = 0$ in D. This proves the theorem.

It should be noted that the above theorem establishes the validity of the Monte Carlo Method for the important case of $\alpha = 0$, $\beta > 0$ in equation (1).

7. Remark

We have carried out the above analysis under the assumption $k_1 \neq k_2$. If $k_1 = k_2 = k$, everything above is valid except that appropriate changes should be made in equations (3) to (7) and the quantities F_1, F_2, F_3 and F_4 in equations (9) should now read

$$F_1 = (I_0(k\,r) + (k\,r/2)\, I_1(k\,r))/I_0^2(k\,r)\,, \tag{19a}$$

$$F_2 = -\,r\, I_1(k\,r)/(2\,k\, I_0^2(k\,r))\,, \tag{19b}$$

$$F_3 = k^3\, r\, I_1(k\,r)/(2\, I_0^2(k\,r))\,, \tag{19c}$$

$$F_4 = (I_0(k\,r) - (k\,r/2)\, I_1(k\,r))/I_0^2(k\,r)\,. \tag{19d}$$

References

1 AGGARWALA, B. D. and IQBAL, M., Limiting Nusselt numbers from membrane analogy for combined free and forced convection through vertical ducts, International Journal of Heat and Mass Transfer **12**, pp. 737–748 (1969).
2 COURANT, R. and HILBERT, D., Methods of Mathematical Physics, vol. 2, Interscience Publishers, (1962).
3 GOPALSAMY, K. and AGGARWALA, B. D., Monte Carlo Methods of some fourth order partial differential equations, ZAMM **50**, pp. 759–767 (1970).
4 GOPALSAMY, K. and AGGARWALA, B. D., On a Monte Carlo method for biharmonic boundary value problems, ZAMM **53**, pp. 293–298 (1973).
5 HIGUCHI, K. and TAKAHASHI, T., Monte Carlo solutions of some boundary value problems of biharmonics, Proceedings of the 12th National Congress of Applied Mechanics, Japan, pp. 197–200 (1962).
6 HAJI-SHEIKH, A. and SPARROW, E. M., The solution of heat conduction problems by probability methods, Trans. ASME, Journal of Heat Transfer **89**, pp. 121–131 (1967).
7 MORLEY, L. S. D., Skew Plates and Structures, Pergamon Press (1963).
8 TIMOSHENKO, S. and WOINOWSKY-KREIGER, S., Theory of Plates and Shells, McGraw-Hill Book Company (1959).
9 WATSON, G. N., A Treatise on the Theory of Bessel Functions, Cambridge University Press (1922).

Eingereicht am 10.11. 1972 (rev. Fassung: 10.8. 1973).

Anschrift: B. D. AGGARWALA, Mathematics Department, University of Calgary, Calgary, Canada

ZAMM 54, 197 – 198 (1974)

G. Baumann

Extrapolation des Series Aleatoires

De nombreux phénomènes physiques ne sont connus que dans un intervalle de temps limité, soit par une fonction $f(t)$, soit par une série chronologique $\sum a_i t_i$ obtenues à partir de mesures expérimentales effectuées dans cet intervalle.

On rencontre ces problèmes dans toute la physique où l'expérience est ramenée à l'étude d'un spectre et particulièrement pour des courbes d'autocorrélation de phénomènes aléatoires et dans la théorie de l'information.

Le problème est d'ailleurs le même pour les séries chronologiques économiques. La prolongation de ces séries permet la prédiction d'avenir des variables économiques. Il est souvent très utile et généralement fort tentant de connaître la variation du phénomène en dehors de l'intervalle de mesure, donc de prolonger les courbes expérimentales en dehors de l'intervalle où l'on a pu effectuer des mesures. Mais des précautions sont nécessaires pour obtenir une prolongation possédant un degré de confiance valable.

On peut montrer que le passage possible entre séries chronologique et spectre de fréquence par transformation de Fourier

$$G(\omega) = \frac{1}{2\pi} \int_{-T/2}^{+T/2} f(t)\, e^{-i\omega t}\, dt$$

et par transformation inverse

$$f(t) = \int_{-\infty}^{+\infty} G(\omega)\, e^{i\omega t}\, d\omega$$

permet l'utilisation de méthodes de prévisions très générales. Nous pouvons donc utiliser indifféremment l'une des deux fonctions dans les méthodes de prédictions:

– prolongement du spectre expérimental en dehors de l'intervalle pour lequel existent des mesures;

– prolongement de la courbe $f(t)$ par extrapolation. On sait que, pour toutes fonctions f entière $\in \mathbb{R}$, une telle extrapolation pourrait être faite en calculant les dérivées successives en un point de l'intervalle où f est connu et en formant de développement en séries de Taylor correspondant (et convergent). Mais comme on part d'une courbe expérimentale, et avec la nécessité d'une coupure de $f(t)$, on obtient une approximation assez mauvaise pour les valeurs élevées de la variable.

Il faut donc trouver une famille de fonctions orthogonales et complètes sur $[-T/2, T/2]$ intervalle où est connu $f(t)$ de façon à pouvoir développer $f(t)$ dans cette base

$$f(t) = \sum_{n=0}^{\infty} a_n \psi_n(t) \tag{1}$$

et que ce développement soit valable à l'intérieur de ce compact.

On peut montrer que ces deux conditions sont remplies par les fonctions sphéroïdales définies par l'équation intégrale

$$\lambda_n \psi_n(t) = \int_{-T/2}^{+T/2} \frac{\sin \Omega (t-u)}{\pi (t-u)} \psi_n(u)\, du\,. \tag{2}$$

Ω étant la plus grande valeur de ω telle que $S(\omega)$, transformée de Fourier de $f(t)$ est non identiquement nulle pour $\omega > \Omega$.

Les fonctions $\psi_n(t)$ sont orthogonales et complètes sur $\mathbb{R}$ et sur $[-T/2, T/2]$. Ce développement est particulièrement intéressant puisqu'il permet de calculer les coefficients a_n par un développement

$$a_n = \frac{1}{\lambda_n} \int_{-T/2}^{+T/2} f(t)\, \psi_n(t)\, dt \tag{3}$$

c'est-à-dire que tous les coefficients sont calculables à partir de la connaissance de $f(t)$ sur l'intervalle $[-T/2, T/2]$ uniquement.

D'après (1) on a

$$a_n = \int_{-\infty}^{+\infty} f(t)\, \psi_n(t)\, dt \quad \text{et} \quad \sum_{n=0}^{\infty} a_n^2 = \int_{-\infty}^{+\infty} f(t)^2\, dt$$

avec

$$\lim_{N\to\infty} \int_{-\infty}^{+\infty} \left[f(t) - \sum_{n=0}^{N} a_n \psi_n(t) \right]^2 dt = 0$$

on peut donc remplacer

$$f(t) \text{ par } f_N(t) = \sum_{n=0}^{N} a_n \psi_n(t)$$

avec

$$a_n = \frac{1}{\lambda_n} \int_{-T/2}^{+T/2} f(t)\, \psi_n(t)\, dt\,.$$

L'intégrale

$$\int_{-\infty}^{+\infty} [f(t) - f_N(t)]^2\, dt = \sum_{n=N+1}^{\infty} a_n^2 \tag{4}$$

convergente en moyenne quadratique devient, dans l'intervalle $[-T/2, T/2]$

$$\int_{-T/2}^{+T/2} [f(t) - f_N(t)]^2\, dt = \sum_{n=N+1}^{\infty} a_n^2 \lambda_n$$

avec λ_n tendant vers zéro quand n augmente, mais pour lesquelles (4) est encore assez grand [2].

On voit donc d'après la définition même qu'on se limite au sous-ensemble B_Ω de $\mathcal{L}^2_\infty$ [$\mathcal{L}^2_\infty$ est l'ensemble des fonctions définies de carré sommable dans $\mathbb{R}$] des fonctions continues de carré sommable sur $\mathbb{R}$ et dont la transformée de Fourier est à support compact. Le problème est donc de voir s'il existe un critère de coupure optimale minimisant l'erreur sur la prédiction.

D'autre part, le développement fini en série de $f(t)$ a une transformée de Fourier à support compact, les $\psi_n(t)$ ayant des transformées de Fourier à support compact, ce qui ne serait pas le cas pour un développement fini en série de Taylor.

Nous allons porter notre étude sur les propriétés de l'opérateur v_Ω défini par

$$B \xrightarrow{v_\Omega} B_\Omega(\mathbb{R})$$
$$f \qquad g = v_\Omega f$$

avec

$$g(t) = \int_{-T/2}^{+T/2} \frac{\sin \Omega (t-u)}{\pi (t-u)} f(u)\, du$$

et de l'opérateur V_Ω défini par

$$B_\Omega \xrightarrow{V_\Omega} B_\Omega(\mathbb{R})$$
$$f \qquad g = V_\Omega f$$

avec

$$g(t) = \int_{-\infty}^{+\infty} \frac{\sin \Omega (t-u)}{\pi (t-u)} f(u)\, du\,.$$

Les transformées $v_\Omega f$ et $V_\Omega f$ appartiennent à B_Ω.

On a

$$v_\Omega f = \int_{-T/2}^{+T/2} N_\Omega (t-u) f(u)\, du\,.$$

$N_\Omega (t-u)$ étant le noyau de l'équation intégrale considérée.

On a un produit de convolutions si on écrit

$$g(t) = \int_{-\infty}^{+\infty} N_\Omega (t-u)\, h(u)\, du$$

avec

$$h(u) = \begin{cases} f(u) \text{ pour } u \in \left[-\frac{T}{2}, \frac{T}{2}\right] \\ 0 \text{ pour } u \notin \left[-\frac{T}{2}, \frac{T}{2}\right] \end{cases}$$

Comme h et $N_\Omega \in \mathcal{L}^2_\infty$, on peut écrire

$$\mathcal{F}(g(t)) = \mathcal{F}(N_\Omega(u)) \, \mathcal{F}(h(u)) \, du$$

$$\mathcal{F}(N_\Omega(u)) = \begin{cases} 1 \quad \text{pour} \quad \omega \in \left[-\frac{T}{2}, \frac{T}{2}\right] \\ 0 \quad \text{pour} \quad \omega \notin \left[-\frac{T}{2}, \frac{T}{2}\right] \end{cases}$$

donc $g(t) \in B_\Omega$.

On peut faire la même démonstration pour $V_\Omega f$, donc $N_\Omega f$ et $V_\Omega f$ opèrent dans B_Ω.

Nous pouvons maintenant montrer que la connaissance de $f(t)$ dans $[-T/2, T/2]$ implique la connaissance de $f(t)$ sur $\mathbb{R}$.

$f(t) \in B_\Omega$, donc on a

$$f(t) = \frac{1}{2\pi} \int_{-\Omega}^{+\Omega} \mathrm{e}^{-i\omega t} \, \mathcal{F}(f(t)) \, d\omega$$

mais $\mathcal{F}(f(t))$ est continu et à support compact, donc bornée.

La dérivée $n^{\text{ième}}$ de $f(t)$ devient donc

$$f^{(n)}(t) = \frac{1}{2\pi} \int_{-\Omega}^{+\Omega} (-i)^n \, \omega^n \, \mathrm{e}^{-i\omega t} \, \mathcal{F}(f(t)) \, d\omega$$

avec $|f^{(n)}(t)| \to M' \, \omega$ donc $\frac{f^n(0)}{n!} \leqq 0$ pour n assez grand.

Le développement en série de TAYLOR de $f(t)$ a donc un rayon de convergence infini, donc la connaissance de $f(t)$ dans $[-T/2, T/2]$ implique la connaissance de $f(t)$ sur $\mathbb{R}$.

References

1 POGORZELSKI, W., Integral Equations and their Applications, Pergamon Press, 1966.
2 SLEPIAN, D., POLLAK, H. O., Prolate Sheroidal Wave Fonctions, Fourier Analysis and Uncertainty, Bell System Technical Journal 1961.

Eingreicht am 1. 2. 1973

Anschrift: Prof. GERMAIN BAUMANN, Université de Nancy, Laboratoir de Physique Théorique, Boulevard des Aiquillettes, Case off. No 140, 54037 Nancy, France

ZAMM 54, 198 (1974)

G. GUDEHUS/I. VARDULAKIS

Darstellung der nichtlinearen Mohrschen Grenzbedingung im Hauptspannungsraum

Die MOHRsche Grenzbedingung besagt bekanntlich, daß die zu Bruchzuständen gehörenden Spannungskreise in der MOHRschen τ, σ-Ebene für ein bestimmtes Material eine gemeinsame Umhüllende $\tau = f(\sigma)$ haben. Diese Bedingung hat die Besonderheit, daß die mittlere Hauptspannung keinen Einfluß hat. Im Raum der Hauptspannungen entspricht die lineare Bedingung $f = c + \sigma \tan \varrho$ einer sechskantigen Pyramidenfläche [1]. Für Erdstoffe, Fels und Beton [2], [3] ist gerade im Bereich höherer Drücke die gemessene MOHRsche Umhüllende nichtlinear. Wir geben die hierzu gehörende Fläche im Hauptspannungsraum an.

Die Umhüllende sei monoton, konvex und habe eine ganz unter der σ-Achse liegende Evolute. Man kann nur dann $f(\sigma)$ umrechnen in eine Grenzkurve $\sigma_i = f^*(\sigma_j)$ in der Ebene der größten und kleinsten Hauptspannungen σ_i und σ_j (Druck positiv). Im $\sigma_1, \sigma_2, \sigma_3$-Raum gehört derjenige Teil des Zylinders $\sigma_i = f^*(\sigma_j)$ zur gesuchten Grenzfläche, für den die dritte Hauptspannung σ_k $\sigma_j \leqq \sigma_k \leqq \sigma_i$ erfüllt (Bild 1). Durch Permutation von i, j, k ergeben sich sechs derartige Zylinder, die zusammen eine zur Raumdiagonale $\sigma_1 = \sigma_2 = \sigma_3$ symmetrische krummlinige sechskantige Pyramide bilden. Die Schnittkurve mit einer Deviatorebene $\sigma_1 + \sigma_2 + \sigma_3 = \text{const}$ ist ein krummliniges, gleichseitiges, aber nicht gleichwinkliges Sechseck (Bild 2).

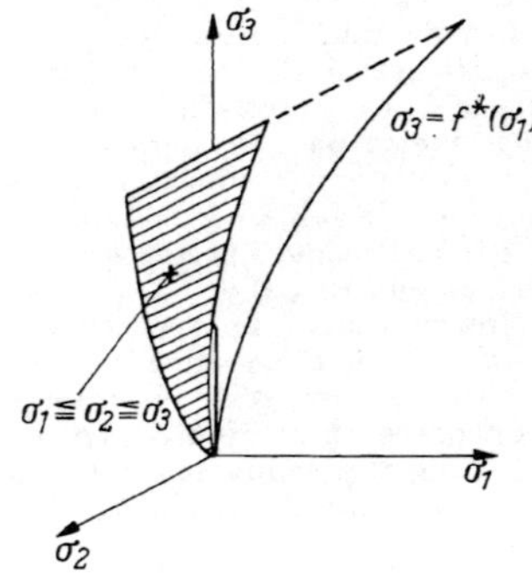

Bild 1

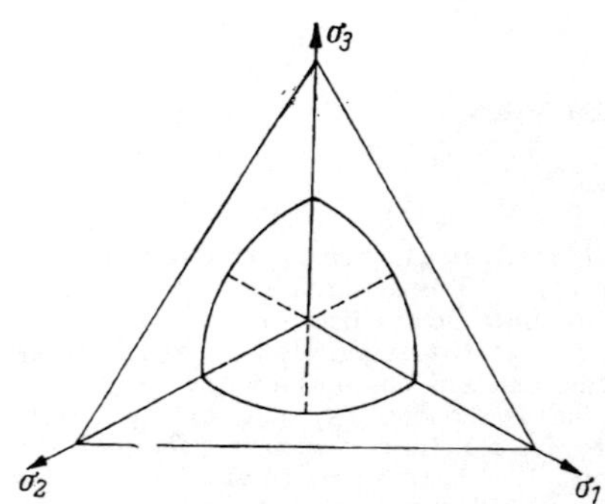

Bild 2

Diese Darstellung ist von Nutzen für die Behandlung räumlicher Bruchprobleme. Man vergleiche z. B. Bild 2 mit Bild 3 in [3] (in letzterem ist versehentlich die Schnittkurve um 60° verdreht gezeichnet).

Literatur

1 SCHELD, R. T., Journ. Mech. Phys. Solids, 4, pp. 10–16 (1955).
2 MÜLLER, L., Der Felsbau, Kap. 2.33; Bd. I, 1, Enke, Stuttgart (1963).
3 RIECKMANN, H.-P., Die Bautechnik 8, p. 268 (1971).

Eingereicht am 6. 2. 1973 (gekürzte Fassung am 7. 6. 1973)

Anschrift: Prof. Dr. Ing. G. GUDEHUS, Institut für Bodenmechanik und Felsmechanik, Universität, 75 Karlsruhe, Richard-Willstätter-Allee, Postfach 6380, BRD

ZAMM 54, 198–200 (1974)

G. NATH

An Approximate Method for the Similar Solution of Nonlinear Boundary Layer Equations for Power-Law Fluids

1. Introduction

The laminar boundary layer flows of power-law non-NEWTONian fluids past external surfaces with or without pressure gradient and mass transfer have been studied by several authors [1]–[3]. The governing nonlinear equations have been solved numerically using RUNGE-KUTTA-GILL method which takes a long computing time.

The present analysis is intended to provide a simple and fast, yet accurate, method for solving nonlinear boundary layer equations governing the flow of power-law non-NEWTONian fluids past an infinite wedge with mass transfer. The important feature of the method is that we have to solve an algebraic equation instead of a nonlinear ordinary differential equation with split two-point boundary conditions. The surface skin friction coefficient obtained by the present method in the presence of suction differs by about 2 to 3 per cent from the corresponding exact results obtained by numerically solving the two-point boundary value problem. The present method can also be applied to other related problems.

2. Boundary layer equations

We consider the steady incompressible laminar viscous flow of a constant property power-law non-NEWTONian fluid past an infinite porous wedge with suction. The governing boundary layer equation can be expressed as [1]–[3].

Continuity:

$$\frac{\partial u}{\partial x} + \frac{\partial v}{\partial y} = 0\,, \tag{1}$$

Momentum:

$$\varrho\, u \frac{\partial u}{\partial x} + \varrho\, v \frac{\partial u}{\partial y} = U \frac{dU}{dx} + K \frac{\partial}{\partial y}\left[\left(\frac{\partial u}{\partial y}\right)^N\right]. \tag{2}$$

The boundary conditions are

$$y = 0:\ u = 0\,, \quad v = v_0\ (v_0 < 0 \text{ for suction})$$

$$y \to \infty:\ u \to U\,, \tag{3}$$

where x and y are the directions along and perpendicular to the body, u and v are the velocity components along x and y directions, ϱ is the density, U is the velocity at the edge of the boundary layer, and K and N are empirically determined constants. It has been assumed [2] that there is no velocity overshoot (i.e. the velocity gradient is always positive).

If the external velocity $U(x)$ varies as x^m and the suction velocity $v_0(x)$ varies as $x^{[m(2N-1)-N]/[N+1]}$, then it is possible to choose a dimensionless coordinate η and a dimensionless stream function $F(\eta)$ so that the continuity equation (1) is identically satisfied and equation (2) is reduced to an ordinary differential equation. The transformations are

$$\eta = \frac{y}{x}\left[\frac{m+1}{N(N+1)} R_x\right]^{1/(N+1)},$$

$$F(\eta) = \left[\frac{m+1}{N(N+1)} R_x\right]^{1/(N+1)} \frac{\psi}{U x}\,, \tag{4}$$

where $R_x = \dfrac{\varrho\, x^N U^{2-N}}{K}$ is the local REYNOLDS number, ψ is the stream function and m is a constant. With the help of (4) equation (2) is reduced to a nonlinear ordinary equation

$$F''' + [\beta (N-1) + 1]\, F (F'')^{2-N} + \beta\left(\frac{N+1}{2}\right)(1 - F'^2)(F'')^{1-N} = 0\,, \tag{5}$$

where $\beta = 2\,m/(m+1)$ is the pressure gradient parameter and prime denotes differentiation with respect to η. The boundary conditions are

$$F(0) = F_w\,, \quad F'(0) = 0\,, \quad F'(\infty) \to 1\,, \tag{6}$$

where F_w is the mass transfer parameter ($F_w > 0$ for suction)

It may be noted that equation (5) reduces to the corresponding boundary layer equation for NEWTONian fluids for $N = 1$. In the present case, we intend to obtain an approximate closed form solution of equation (5) using an iterative procedure. Following WEYL [4], we introduce a sequence of functions $\{F_n(\eta)\}$ which satisfy

$$F_n''' + [\beta(N-1)+1]\, F_{n-1} (F_{n-1}'')^{2-N} + \beta\left(\frac{N+1}{2}\right)(1 - F_{n-1}'^2)(F_{n-1}'')^{1-N} = 0\,. \tag{7}$$

The boundary conditions (6) can be written as

$$F_n(0) = F_w\,, \quad F_n'(0) = 0\,, \quad F_n'(\infty) \to 1\,, \tag{8}$$

where the subscript n denotes n^{th} iteration.

3. Solution of boundary layer equations

We integrate equation (7) with respect to η between the limits 0 and η

$$F_n''(\eta) - F_n''(0) = -[\beta(N-1)+1] \int_0^\eta F_{n-1} (F_{n-1}'')^{2-N}\, d\eta - \beta\left(\frac{N+1}{2}\right)\int_0^\eta (1 - F_{n-1}'^2)(F_{n-1}'')^{1-N}\, d\eta\,. \tag{9}$$

We can solve equation (9) by iteration by introducing a suitable first approximation $F_0''(\eta)$ (and hence $F_0(\eta)$ and $F_0'(\eta)$) in the right hand side of equation (9). Consequently, we obtain $F_1''(0)$ or $F_1''(\eta)$ (using the appropriate boundary conditions) which is the second approximation. The shear stress function $F_1''(\eta)$ thus obtained is further integrated to determine the unknown constant present in $F_1''(\eta)$ using the condition that $F_1'(\infty) \to 1$. Then we calculate $F_1''(0)$ for different values of the parameters β, F_w, and N to see how closely the results compare with those of exact results. It may be remarked that in the present analysis the results of the second approximation are in close agreement with the exact results, hence it is not necessary to go beyond second approximation (i.e., beyond first iteration).

We know that the shear stress function $F_0''(\eta)$ has a finite value at $\eta = 0$ and it tends to zero as η tends to infinity. We assume an expression for $F_0''(\eta)$ satisfying these conditions

$$F_0''(\eta) = \lambda \exp(-\lambda\,\eta) \tag{10}$$

where λ ($\lambda > 0$) is a constant to be determined. Hence $F_0'(\eta)$ and $F_0(\eta)$ under conditions (8) are given by

$$F_0'(\eta) = 1 - \exp(-\lambda\,\eta)\,, \tag{11}$$

$$F_0(\eta) = F_w + \eta - \frac{1}{\lambda}[1 - \exp(-\lambda\,\eta)]\,. \tag{12}$$

It may be noted that the assumed profile form will give good representation for full profiles without velocity overshoot, but will fail for separating profiles. As the present analysis is intended to study flow problems in the absence of velocity overshoot and separation, the present profile form serves our purpose.

We substitute for $F_0(\eta)$, $F_0'(\eta)$, and $F_0''(\eta)$ from equations (10) to (12) in equation (9) and use the condition that $F_0''(\infty) \to 0$. As a result, we get an expression for $F_0''(0)$ (second approximation) in the form

$$F_1''(0) = \lambda^{-N}\left[\{\beta(N-1)+1\}\left\{\frac{(\lambda F_w - 1)}{(2-N)} + \frac{1}{(2-N)^2} + \frac{1}{(3-N)}\right\} + \frac{\beta(N+1)(4-N)}{2(2-N)(3-N)}\right]. \tag{13}$$

Integrating equation (7) between the limits η and ∞ and using the condition that $F_1''(\infty) \to 0$, we get $F_1''(\eta)$ in the form

$$F_1''(\eta) = \left[\{\beta(N-1)+1\}\left\{\frac{\lambda(F_w+\eta) - 1}{2-N} + \frac{1}{(2-N)^2}\right\} + \frac{\beta(N+1)}{(2-N)}\right]\lambda^{-N}\exp\{-(2-N)\lambda\,\eta\} + \left[\frac{\beta(N-3)+2}{2(3-N)}\right]\lambda^{-N}\exp\{-(3-N)\lambda\,\eta\}\,. \tag{14}$$

Again integrating equation (14) between the limits 0 and ∞ and using the condition that $F_1'(\infty) \to 1$, we get

$$\lambda^{1+N} = [\beta(N-1)+1]\left[\frac{\lambda F_w - 1}{(2-N)^2} + \frac{1}{(3-N)^2} + \frac{2}{(2-N)^3}\right] + \beta(N+1)\left[\frac{1}{(2-N)^2} - \frac{1}{2(3-N)^2}\right]. \tag{15}$$

The solution of equation (15) for prescribed values of F_w, β and N can be obtained numerically by NEWTON-RAPHSON method. Knowing λ, $F_1''(0)$ can be obtained from equation (13).

The local skin friction coefficient can be expressed as

$$C_f = \frac{\tau}{\frac{1}{2}\varrho\, U^2} = C(N, \beta, F_w)\, R_x^{-1/(N+1)}\ , \tag{16}$$

where

$$C(N, \beta, F_w) = 2\left[\frac{2}{N\,(N+1)\,(2-\beta)}\right]^{N/(N+1)} [F_1''(0)]^N \tag{17}$$

and

$$\tau = K\left|\frac{\partial u}{\partial y}\right|^{N-1}\frac{\partial u}{\partial y} = K\left(\frac{\partial u}{\partial y}\right)^N$$

is the shear stress [2]. It may be remarked that the corresponding expressions for $F_1''(0)$, λ, and C_f for the case of NEWTONian Fluids can be obtained by putting $N = 1$ in equations (13), (15)–(17).

4. Numerical results and discussion

The governing algebraic equation (15) has been solved numerically using NEWTON-RAPHSON iterative procedure on IBM 360 digital computer for a range of values ($0.1 \leqq N \leqq 1.0$) of the exponent of the power law shear model and a range of values ($0.5 \leqq F_w \leqq 2.0$) of the dimensionless suction parameter. The pressure gradient parameter β was varied from 0 to 1. The results along with the exact results (obtained by numerically solving the equation (5) under the conditions (6) using fourth order RUNGE-KUTTA-GILL algorithm) have been shown in Figs. 1–4.

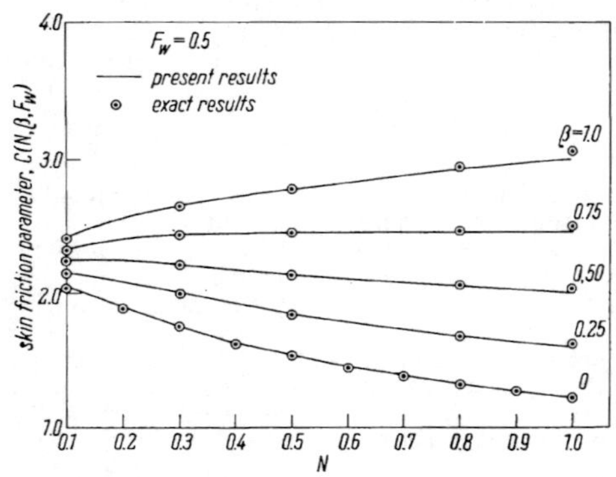

Fig. 1. Variation of the skin friction parameter $C(N, \beta, F_w)$ with N for $F_w = 0.5$

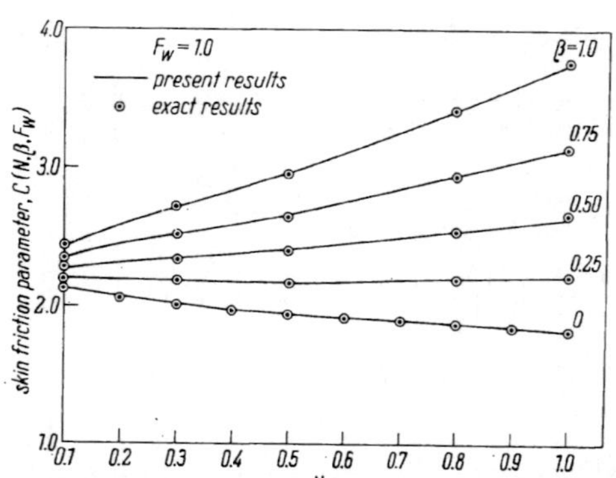

Fig. 2. Variation of the skin friction parameter $C(N, \beta, F_w)$ with N for $F_w = 1.0$

The results as shown in Figs. 1–4, indicate that $C(N, \beta, F_w)$, for a given N, increases with F_w and β. However, for certain F_w ($0.5 \leqq F_w \leqq 1.0$) and β ($0 \leqq \beta \leqq 0.5$), it decreases with N. It is important to note that the present results differ from the exact results by about 2 to 3 per cent and the total computer time taken is about 2 minutes as compared to about 3 hours required for the numerical solution of the corresponding two-point boundary value problem using RUNGE-KUTTA-GILL algorithm for the same range of parameters. It may be remarked that the results of the second approximation in the present iterative procedure are in close agreement with the exact numerical results, thus indicating the convergence of the procedure. The detailed discussion on the convergence of the procedure will be given in later publications.

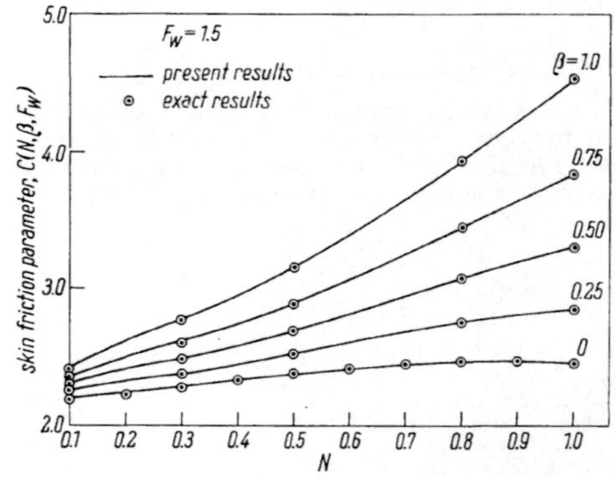

Fig. 3. Variation of the skin friction parameter $C(N, \beta, F_w)$ with N for $F_w = 1.5$

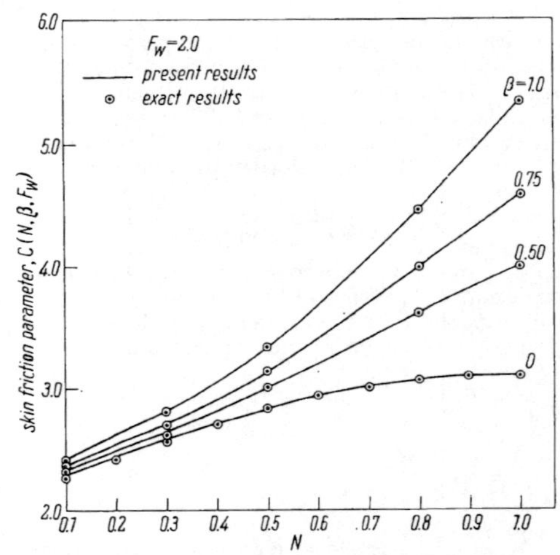

Fig. 4. Variation of the skin friction parameter $C(N, \beta, F_w)$ with N for $F_w = 2.0$

5. Conclusions

The present analysis provides a simple and fast, yet quite accurate, method for solving the boundary layer equations governing the flow of power-law fluids past a wedge. The results obtained by the present method differ from the corresponding exact results by about 2 to 3 per cent.

References

1 KAPUR, J. N. and SRIVASTAVA, R. C., Similar Solutions of the Boundary Layer Equations for Power-Law Fluids, ZAMP **14**, pp. 383–388 (1963).
2 THOMPSON, E. R. and SNYDER, W. T., Laminar Boundary Layer Flows of Power-Law Non-Newtonian Fluids Past External Surfaces with Mass and Heat Transfer, Proc. 5th Intl. Congr. Rheology, Vol. **1**, Ed. S. Onogi, Tokyo Press, Tokyo, 1969.
3 HAYASI, N., Similarity of the Two-Dimensional and Axisymmetric Boundary Layer Flows for Purely Viscous Non-Newtonian Fluids, J. Fluid Mech. **23**, pp. 293–303 (1965).
4 WEYL, W., Concerning the Differential Equations of Some Boundary Layer Problems, Proc. Nat. Acad. Sci., Wash., **27**, pp. 578–583 (1941).

Eingereicht am 14. 3. 1973
(gekürzte Fassung am 9. 6. 1973)

Anschrift: Dr. GIRISHWAR NATH, Department of Applied Mathematics, Indian Institute of Science, Bangalore 560012, India

ZAMM 54, 201 – 203 (1974)

B. M. SINGH

Quadruple Integral Equations Involving Inverse Mellin Transforms

1. Introduction

Some dual integral equations involving inverse MELLIN transforms were first considered by SRIVASTAV [1], which have been shown to be reducible to FREDHOLM integral equations. Later on SRIVASTAV and PARIHAR [2] presented closed form solutions of the same dual integral equations involving inverse MELLIN transforms. They also reduced certain triple integral equations to dual series equations whose closed form solution is known. Recently SINGH [4] obtained an exact solution of triple integral equations involving inverse MELLIN transforms by finite HILBERT transform technique. The advantage of the above mentioned technique is that the solution obtained is simpler than that given by SRIVASTAV and PARIHAR [2]. This paper shall be devoted to finding the solution of quadruple integral equations involving inverse MELLIN Transforms. These integral equations which arise in four-part mixed boundary value problems for wedge-shaped regions. We prefer to reduce the quadruple integral equations to a singular integral equation, which is solvable in a closed form. No attempt is made here to justify the change of order of integrations and the analysis is quite formal.

In the last section of this paper an application of quadruple integral equations to a two-dimensional electrostatic problem has also been discussed.

2. Quadruple integral equations

Let us consider the quadruple integral equations:

$$\frac{1}{2\pi i}\int_{c-i\infty}^{c+i\infty}\psi(s)\tan\alpha s\,\varrho^{-s}\,ds=f_1(\varrho)\,,\qquad 0<\varrho<a\,,\tag{1}$$

$$\frac{1}{2\pi i}\int_{c-i\infty}^{c+i\infty}s\,\psi(s)\,\varrho^{-s}\,ds=f_2(\varrho)\,,\qquad a<\varrho<b\,,\tag{2}$$

$$\frac{1}{2\pi i}\int_{c-i\infty}^{c+i\infty}\psi(s)\tan\alpha s\,\varrho^{-s}\,ds=f_3(\varrho)\,,\qquad b<\varrho<1\,,\tag{3}$$

$$\frac{1}{2\pi i}\int_{c-i\infty}^{c+i\infty}s\,\psi(s)\,\varrho^{-s}\,ds=f_4(\varrho)\,,\qquad \varrho>1\,.\tag{4}$$

We assume in the first instance that $f_3=0, f_4=0$. Equation (4) is identically satisfied if we take (Cf. SRIVASTAV and PARIHAR [2], p. 131)

$$\psi(s)=h^{-1}\beta\left(\frac{1}{2},\frac{s}{h}\right)\int_0^1 t^{s-\frac{h}{2}}\,g(t^h)\,dt\,,\tag{5}$$

$$h=\pi/\alpha\,,\qquad h>0\,,$$

where $g(t^h)$ is an auxiliary function to be determined from the conditions (1), (2) and (3). Substituting for $\psi(s)$ in (1), (2) and (3), then on evaluating the complex integrals we get the triple integral equations:

$$\int_0^\varrho g(t^h)\,(\varrho^h-t^h)^{-1/2}\,dt=f_1(\varrho)\,,\qquad 0<\varrho<a\,,\tag{6}$$

$$\frac{\partial}{\partial\varrho}\int_\varrho^1 g(t^h)\,(t^h-\varrho^h)^{-1/2}\,dt=\frac{f_2(\varrho)}{\varrho}\,,\qquad a<\varrho<b\,,\tag{7}$$

$$\int_0^\varrho g(t^h)\,(\varrho^h-t^h)^{-1/2}\,dt=0\,,\qquad b<\varrho<1\,.\tag{8}$$

To solve the above triple integral equations we assume that:

$$\frac{\partial}{\partial\varrho}\int_0^\varrho\frac{g(t^h)\,dt}{(\varrho^h-t^h)^{1/2}}=R(\varrho^h)\,,\qquad a<\varrho<1\,,\tag{9}$$

where the function $R(\varrho^h)$ is to be determined. Differentiating (6) with respect to ϱ and then by making use of the solution of ABEL's type integral equation from the book of SNEDDON ([3], pp. 40 – 41) we get from (6) and (9) that

$$g(t^h)=\frac{h}{\pi}\,t^{h-1}\left[\int_0^a\frac{f_1'(\varrho)\,d\varrho}{(t^h-\varrho^h)^{1/2}}+\int_a^t\frac{R(\varrho^h)\,d\varrho}{(t^h-\varrho^h)^{1/2}}\right],\tag{10}$$

$$a<t<1\,.$$

Substituting the value of $g(t^h)$ from (10) in (7) and interchanging the order of integrations we get

$$\frac{1}{\pi}\frac{\partial}{\partial\varrho}\int_a^1 h_2(x^h)\log\left|\frac{\sqrt{1-\varrho^h}+\sqrt{1-x^h}}{\sqrt{1-\varrho^h}-\sqrt{1-x^h}}\right|dx=\frac{f_2(\varrho)}{\varrho}-$$

$$-\frac{1}{\pi}\frac{\partial}{\partial\varrho}\int_0^a f_1'(x)\log\left|\frac{\sqrt{1-\varrho^h}+\sqrt{1-x^h}}{\sqrt{1-x^h}-\sqrt{1-\varrho^h}}\right|dx\,,\qquad a<\varrho<b\,.\tag{11-A}$$

We now differentiate (8) with respect to ϱ, then we have with the help of (9)

$$R(x^h)=\begin{cases}h_2(x^h)\,, & a<x<b\,,\\ 0\,, & b<x<1\,,\end{cases}\tag{11-B}$$

where $h_2(x^h)$ is unknown function to be determined. Hence (11-A) can be written by making use of the above equation

$$\frac{1}{\pi}\frac{\partial}{\partial\varrho}\int_a^b h_2(x^h)\log\left|\frac{\sqrt{1-\varrho^h}+\sqrt{1-x^h}}{\sqrt{1-\varrho^h}-\sqrt{1-x^h}}\right|dx$$

$$=\frac{f_2(\varrho)}{\varrho}-\frac{1}{\pi}\frac{\partial}{\partial\varrho}\int_0^a f_1'(x)\log\left|\frac{\sqrt{1-\varrho^h}+\sqrt{1-x^h}}{\sqrt{1-\varrho^h}-\sqrt{1-x^h}}\right|dx\,,$$

$$a<\varrho<b\,.$$

We find, after interchanging the order of differentiation and integration, $h_2(x^h)$ must satisfy the integral equation

$$\frac{h}{\pi}\int_a^b\frac{h_2(x^h)\sqrt{1-x^h}}{x^h-\varrho^h}\,dx=\frac{\sqrt{1-\varrho^h}}{\varrho^h}\,f_2(\varrho)-$$

$$-\frac{h}{\pi}\int_0^a\frac{f_1'(x)\sqrt{1-x^h}\,dx}{x^h-\varrho^h}=m(\varrho)\;(s\;a\;y)\,,\qquad a<\varrho<b\,.\tag{12}$$

Making use of modified HILBERT transform theorem given by SINGH [4], we obtain the solution of the integral equation (12) in the following form

$$\frac{h_2(x^h)\sqrt{1-x^h}}{x^{h-1}}=-\frac{h}{\pi}\left(\frac{x^h-a^h}{b^h-x^h}\right)^{1/2}\int_a^b\left(\frac{b^h-\varrho^h}{\varrho^h-a^h}\right)^{1/2}\times$$

$$\times\frac{\varrho^{h-1}\,m(\varrho)\,d\varrho}{\varrho^h-x^h}+\frac{c}{[(x^h-a^h)\,(b^h-x^h)]^{1/2}}\,,\qquad a<x<b\,,\tag{13}$$

where c is an arbitrary constant. Making use of (6), (9) and (11-B) we find the expression for $g(t^h)$ in the following form with the help of the solution of ABEL's type integral equation,

$$g(t^h)=\frac{h}{\pi}\,t^{h-1}\left[\int_0^t\frac{G(u^h)\,du}{(t^k-u^h)^{1/2}}\right],\qquad 0<t<1\,,\tag{14}$$

where

$$G(u^h)=\begin{cases}f_1'(u)\,, & 0<u<a\,,\\ h_2(u^h)\,, & a<u<b\,,\\ 0\,, & b<u<1\,.\end{cases}\tag{15}$$

Substituting the expression for $g(t^h)$ from (14) in (8) and interchanging the order of integrations and using the well known integral

$$\int_u^\varrho \frac{t^{h-1}\,dt}{(t^h-u^h)^{1/2}\,(\varrho^h-t^h)^{1/2}} = \frac{\pi}{h}\,, \tag{16}$$

we get

$$\int_0^\varrho G(u^h)\,du = 0\,, \qquad 0<\varrho<1\,. \tag{17}$$

Making use of (15) we get

$$\int_0^a f_1'(u)\,du + \int_a^b h_2(u^h)\,du = 0\,. \tag{18}$$

With the help of (13) we find that

$$c = \frac{h^2\sqrt{1-a^h}}{2\pi F}\int_a^b \frac{x^{h-1}}{\sqrt{1-x^h}}\left(\frac{x^h-a^h}{b^h-x^h}\right)^{1/2} dx \times \\ \times \int_a^b \left(\frac{b^h-\varrho^h}{\varrho^h-a^h}\right)^{1/2} \frac{\varrho^{h-1}\,m(\varrho)\,d\varrho}{\varrho^h-x^h} - \frac{h\sqrt{1-a^h}}{2F}\int_0^a f_1'(u)\,du\,, \tag{19}$$

where

$$F = F\left(\frac{\pi}{2}, \sqrt{\frac{b^h-a^h}{1-a^h}}\right),$$

denotes elliptic integral of first kind.

Equations (5), (10), (11-B), (13) and (19) are sufficient to compute the solution of the system of equations (1), (2), (3) and (4) for $f_3=0, f_4=0$. To complete the solution, assume that $f_1=0, f_2=0$ and $f_3 \neq 0, f_4 \neq 0$.

Let

$$\psi(s) = h^{-1}\beta\left(\frac{1}{2}, \frac{s}{h}\right)\int_a^\infty t^{s-h/2}\,g_2(t^h)\,dt\,, \tag{20}$$

where $g_2(t^h)$ is an auxiliary function to determined. Making use of the paper of SRIVASTAV and PARIHAR [1] we find that equation (1) is automatically satisfied by taking the values of $\psi(s)$ given by (20). Substituting for $\psi(s)$ in (2), (3) and (4) we obtain the triple integral equations

$$\frac{\partial}{\partial\varrho}\int_\varrho^\infty \frac{g_2(t^h)\,dt}{(t^h-\varrho^h)^{1/2}} = 0\,, \qquad a<\varrho<b\,, \tag{21}$$

$$\int_a^\varrho \frac{g_2(t^h)\,dt}{(\varrho^h-t^h)^{1/2}} = f_3(\varrho)\,, \qquad b<\varrho<1\,, \tag{22}$$

$$\frac{\partial}{\partial\varrho}\int_\varrho^\infty \frac{g_2(t^h)\,dt}{(t^h-\varrho^h)^{1/2}} = \frac{-f_4(\varrho)}{\varrho}\,, \quad 1<\varrho\,. \tag{23}$$

Further let

$$\frac{\partial}{\partial\varrho}\int_\varrho^\infty \frac{g_2(t^h)\,dt}{\sqrt{t^h-\varrho^h}} = -R_2(\varrho^h)\,, \qquad a<\varrho<1\,, \tag{24}$$

where $R_2(\varrho^h)$ is to be determined. With the help of the solution of ABEL's type integral equation given in the book of SNEDDON ([3], p. 41) we get from (23) and (24) that

$$g_2(t^h) = \frac{h}{\pi}t^{h-1}\left[\int_t^1 \frac{R_2(u^h)\,du}{(u^h-t^h)^{1/2}} + \int_1^\infty \frac{f_4(u)\,du}{u\,(u^h-t^h)^{1/2}}\right], \\ a<t<1\,. \tag{25}$$

We have that for $b<\varrho<1$

$$\int_a^\varrho \frac{t^{h-1}\,dt}{\sqrt{\varrho^h-t^h}}\int_t^1 \frac{R_2(u^h)\,du}{\sqrt{u^h-t^h}} = \int_a^\varrho\left[\int_t^\varrho + \int_\varrho^1\right] t^{h-1}\,R(u^h) \times \\ \times \frac{du\,dt}{\sqrt{\varrho^h-t^h}\sqrt{u^h-t^h}} = \left[\int_a^\varrho\int_a^u + \int_\varrho^1\int_a^\varrho\right]\frac{t^{h-1}\,R_2(u^h)\,dt\,du}{\sqrt{\varrho^h-t^h}\sqrt{u^h-t^h}} \\ = \int_a^1 R_2(u^h)\,du\int_a^{\min(u,\varrho)} \frac{t^{h-1}\,dt}{\sqrt{\varrho^h-t^h}\sqrt{u^h-t^h}} \\ = \frac{1}{h}\int_a^1 R_2(u^h)\log\left|\frac{\sqrt{u^h-a^h}+\sqrt{\varrho^h-a^h}}{\sqrt{u^h-a^h}-\sqrt{\varrho^h-a^h}}\right| du\,. \tag{26}$$

Substituting the expression for $g_2(t^h)$ from (25) into (22) and interchanging the order of integrations and using the above result we can easily show that

$$\frac{1}{\pi}\int_a^1 R_2(u^h)\log\left|\frac{\sqrt{u^h-a^h}+\sqrt{\varrho^h-a^h}}{\sqrt{u^h-a^h}-\sqrt{\varrho^h-a^h}}\right| du \\ = f_3(\varrho) - \frac{1}{\pi}\int_1^\infty \frac{f_4(u)}{u}\log\left|\frac{\sqrt{u^h-a^h}+\sqrt{\varrho^h-a^h}}{\sqrt{u^h-a^h}-\sqrt{\varrho^h-a^h}}\right| du\,, \\ b<\varrho<1\,. \tag{27}$$

We can write by making use of (21) and (24)

$$R_2(\varrho^h) = \begin{cases} 0\,, & a<\varrho<b\,, \\ h_3(\varrho^h)\,, & b<\varrho<1\,. \end{cases} \tag{28}$$

Hence with the help of (27) we find that

$$\frac{1}{\pi}\int_b^1 h_3(u^h)\log\left|\frac{\sqrt{u^h-a^h}+\sqrt{\varrho^h-a^h}}{\sqrt{u^h-a^h}-\sqrt{\varrho^h-a^h}}\right| du = f_3(\varrho) - \\ -\frac{1}{\pi}\int_1^\infty \frac{f_4(u)}{u}\log\left|\frac{\sqrt{u^h-a^h}+\sqrt{\varrho^h-a^h}}{\sqrt{u^h-a^h}-\sqrt{\varrho^h-a^h}}\right| du = \pi N(\varrho)\,, \\ \text{(say)},\ b<\varrho<1\,. \tag{29}$$

Making use of the paper of PARIHAR ([5], Lemma 2, p. 257) we get the solution of the integral equation (29) in the following form

$$h_3(u^h) = \frac{h\,u^{h-1}}{\sqrt{(u^h-a^h)(u^h-b^h)(1-u^h)}} \times \\ \times\left[\int_b^1 \frac{\sqrt{(x^h-a^h)(x^h-b^h)(1-x^h)}}{u^h-x^h}N'(x)\,dx + \\ + \frac{B\sqrt{1-a^h}}{4F\left[\frac{\pi}{2}, \sqrt{\frac{1-b^h}{1-a^h}}\right]}\right], \quad b<u<1\,, \tag{30}$$

where

$$B = \frac{\pi h\sqrt{1-a^h}}{F\left[\frac{\pi}{2}, \sqrt{\frac{b^h-a^h}{1-a^h}}\right]}\int_b^1 \frac{x^{h-1}\,N(x)\,dx}{\sqrt{(x^h-a^h)(x^h-b^h)(1-x^h)}} - \\ -2h\int_b^1 \frac{y^{h-1}\,dy}{\sqrt{(y^h-a^h)(y^h-b^h)(1-y^h)}} \times \\ \times\int_b^1 \frac{\sqrt{(x^h-a^h)(x^h-b^h)(1-x^h)}}{y^h-x^h}N'(x)\,dx\,. \tag{31}$$

By combining the two solutions, viz., one for $f_1 = 0, f_2 = 0$, and the other for $f_3 = 0$ $f_4 = 0$ we obtain the solution of the quadruple integral equations (1), (2), (3) and (4). The special case of much practical importance shall be one in which $f_1 = f_2 = f_4 = 0$ and $f_3 = q_0$, where q_0 is a constant. In this case

$$h_3(\varrho^h) = \frac{h\sqrt{1-a^h}\,\varrho^{h-1}\,q_0}{2\sqrt{(\varrho^h - a^h)(q^h - b^h)(1-\varrho^h)}} \times$$

$$\times \frac{1}{K\left(\sqrt{\dfrac{b^h - a^h}{1-a^h}}\right)}, \qquad b < \varrho < 1 . \tag{32}$$

where

$$K\left(\sqrt{\frac{b^h - a^h}{1-a^h}}\right) = F\left[\frac{\pi}{2}, \sqrt{\frac{b^h - a^h}{1-a^h}}\right].$$

Denotes the complete elliptic integral.

3. Two-dimensional electrostatic problem

To illustrate the use of the solution of the quadruple integral equations we consider the wedge-shaped region of angle α, earthed on the face $\theta = \alpha$ and the two strips are lying on the face $\theta = 0$ and in the intervals $0 \leqq \varrho \leqq a$, $b \leqq \varrho \leqq 1$. The strip defined by $b \leqq \varrho \leqq 1$, $\theta = 0$, charged at a constant potential q_0 and the strip $0 \leqq \varrho \leqq a$, $\theta = 0$ is earthed. The face $\theta = 0$ is insulated for $a < \varrho < b$ and $1 < \varrho < \infty$.

We set the following conditions on the potential function V:

$$\nabla^2 V(\varrho, \theta) = \frac{\partial^2 V}{\partial \varrho^2} + \frac{1}{\varrho}\frac{\partial V}{\partial \varrho} + \frac{1}{\varrho^2}\frac{\partial^2 V}{\partial \theta^2} = 0 ,$$

$$\varrho \in (0, \infty) , \qquad 0 < \theta < \alpha , \tag{33}$$

$$V(\varrho, \alpha) = 0 , \qquad 0 < \varrho < \infty , \tag{34}$$

$$\left(\frac{1}{\varrho}\frac{\partial V}{\partial \theta}\right)_{\theta=0} = 0 , \qquad a < \varrho < b , \qquad 1 < \varrho < \infty , \tag{35}$$

$$V(\varrho, 0) = q_0 \,(\text{constant}) , \qquad b < \varrho < 1 , \tag{36}$$

$$V(\varrho, 0) = 0 , \qquad 0 < \varrho < a . \tag{37}$$

The solution of (33), which satisfies the condition (34) has the form:

$$V(\varrho, \theta) = \frac{1}{2\pi i}\int_{c-i\infty}^{c+i\infty} A(s)\sin[(\alpha - \theta)\,s]\,\varrho^{-s}\,ds . \tag{38}$$

The conditions (35), (36) and (37) reduce to the following quadruple integral equations:

$$\frac{1}{2\pi i}\int_{c-i\infty}^{c+i\infty} \psi(s)\tan\alpha s\,\varrho^{-s}\,ds = \begin{cases} 0 , & 0 < \varrho < a , \\ q_0 , & b < \varrho < 1 , \end{cases} \tag{39}$$

$$\frac{1}{2\pi i}\int_{c-i\infty}^{c+i\infty} s\,\psi(s)\,\varrho^{-s}\,ds = \begin{cases} 0 , & a < \varrho < b , \\ 0 , & 1 < \varrho , \end{cases} \tag{40}$$

where

$$\psi(s) = A(s)\cos\alpha\,s .$$

We shall now find the surface charge density of the strip which is equal to:

$$\sigma(\varrho) = -\frac{1}{4\pi\varrho}\left(\frac{\partial V}{\partial \theta}\right)_{\theta=0}$$

$$= \frac{1}{4\pi\varrho}\left[\frac{1}{2\pi i}\int_{c-i\infty}^{c+i\infty} s\,\psi(s)\,\varrho^{-s}\,ds\right], \qquad b < \varrho < 1 . \tag{41}$$

We can now easily show with the help of the section 2 that:

$$\sigma(\varrho) = \frac{h_3(\varrho^h)}{4\pi} = \frac{\varrho^{h-1}\,q_0}{\sqrt{(\varrho^h - a^h)(\varrho^h - b^h)(1-\varrho^h)}} \times$$

$$\times \frac{h\sqrt{1-a^h}}{8\pi\,\mathrm{K}\left(\sqrt{\dfrac{b^h - a^h}{1-a^h}}\right)}, \qquad b < \varrho < 1 . \tag{42}$$

References

1 SRIVASTAV, R. P., Certain two dimensional mixed boundary value problems for wedge-shaped regions and dual integral equations, Proc. Edinburgh Math. Soc. **14**, pp. 321–332 (1965).
2 SRIVASTAV, R. P. and PARIHAR, K. S., Dual and triple integral equations involving inverse Mellin transforms, SIAM J. Appl. Math. **16**, pp. 126–133 (1968).
3 SNEDDON, I. N., Mixed Boundary Value Problems in Potential Theory, John Wiley, New York 1966.
4 SINGH, B. M., Triple integral equations involving inverse Mellin transforms, Defence Science Journal **22**, pp. 153–158 (1972).
5 PARIHAR, K. S., Some Triple Trigonometrical Series Equations and their application, Proc. Roy. Soc. Edin. (A), vol. **69**, pp. 255–265 (1971).

Eingereicht am 30. 3. 1973

Anschrift: Dr. BRIJ MOHAN SINGH, 30/3 Gandhi Colony, Muzaffarnagar (U.P.), India

ZAMM 54, 203–205 (1974)

A. M. LOKOSCHTSCHENKO

Bestimmung der zum Plattdrücken einer Zylinderschale unter konstantem Außendruck erforderlichen Zeit

Näherungsweise wird der Verformungsprozeß eines Ringes unter der Wirkung eines konstanten Außendruckes q im Zustand des Kriechens untersucht. Es wird vorausgesetzt, daß ein Ring der Dicke $2\,h$ und der Breite 1 eine Form hat, die leicht von der Kreisform abweicht. Mit dem angegebenen Berechnungsverfahren wird dann das Verhalten einer langen Zylinderschale für den Fall untersucht, daß Randeinflüsse vernachlässigt werden können. Die Schale befindet sich in einem instationären Zustand; bei der Lösung wird die Verformbarkeit der Mittellinie berücksichtigt. Wie in [1] wird vorausgesetzt, daß während des gesamten Vorgangs der Ring zwei Symmetrieachsen besitzt und als Grundparameter der größte und kleinste Ringdurchmesser $2\,a_1$ und $2\,a_2$ benutzt werden können. Es soll die Zeit berechnet werden, bei der der kleinste Durchmesser null wird.

Neben der üblicherweise getroffenen Annahme vom Ebenbleiben der Querschnitte wird vorausgesetzt, daß der Ring zu einem beliebigen Zeitpunkt durch zwei ineinander übergehende Kreise mit den Radien R_1 und R_2 angenähert werden kann (Bild 1, das Bild zeigt nur den ersten Quadranten).

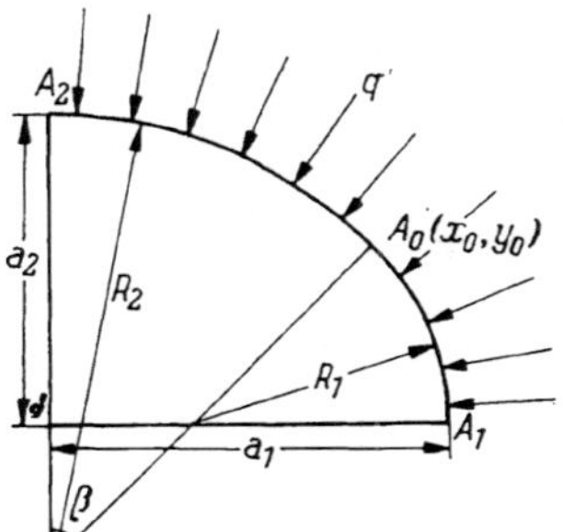

Bild 1

Es soll nun das zeitabhängige Verhalten eines Ringes untersucht werden, der aus einem sich verfestigenden Material besteht, dessen Eigenschaften durch folgende Beziehung zwischen der Spannung σ, der Kriechdeformation p und der Verformungsgeschwindigkeit $\dot{p}$ beschrieben werden kann:

$$\dot{p} = B\,p^{-\alpha}\,\sigma^n , \tag{1}$$

wobei B, α und n Materialkonstanten sind. Der Einfachheit halber soll angenommen werden, daß n als Verhältnis zweier ganzer ungerader Zahlen und α als Verhältnis einer geraden zu einer ungeraden Zahl darstellbar ist. Unter z soll der Abstand eines beliebigen Punktes von der Mittellinie des Ringes, bezogen auf die halbe Dicke verstanden werden. Mit (1) und der Annahme vom Ebenbleiben der Querschnitte erhält man dann:

$$\sigma = (\dot{p}\, p^{\alpha}/B)^{\gamma} = [(\dot{\varepsilon}_0 - \dot{\varkappa}\, h\, z)(\varepsilon_0 - \varkappa\, h\, z)^{\alpha}/B]^{\gamma}, \quad \gamma\, n = 1, \quad (2)$$

wobei ε_0 und $\varkappa$ die Verformung und Krümmung der Mittellinie sind, und der Punkt die Differentiation nach der Zeit t bedeutet. Unter Benutzung der Beziehung (2) und nach Aufstellen der Gleichgewichtsbedingungen in allgemeiner Form erhält man ein System von nichtlinearen Differentialgleichungen, das nur einer numerischen Lösung zugänglich ist. Zur Lösung der Aufgabe soll deshalb eine Näherungsmethode angegeben werden. Der Einfachheit halber soll angenommen werden, daß sich die zeitliche Änderung der Mittelliniendeformation und der Krümmung nach ein und demselben Gesetz vollzieht. Mit anderen Worten, es wird vorausgesetzt, daß eine Zeitfunktion $s(t)$ existiert (die sich von Punkt zu Punkt ändert), mit deren Hilfe man die Verformung und die Krümmung in jedem Punkt der Mittellinie aus deren Geschwindigkeiten ermitteln kann:

$$\varepsilon_0 = \dot{\varepsilon}_0\, s(t), \qquad \varkappa = \dot{\varkappa}\, s(t).$$

Dabei ist in beiden Beziehungen $s(t)$ ein und dieselbe Funktion.

Untersuchen wir nun das Verhalten des Ringes analog zur Behandlung der Knickung eines visko-plastischen Stabes [2]. Ist y das Verhältnis $\dot{\varepsilon}_0/(h\,\dot{\varkappa})$, so erhält man aus (2):

$$\sigma = B^{-\gamma}\, s^{\alpha\gamma} (h\,\dot{\varkappa})^{(k-1)} (y - z)^{(k-1)}, \qquad k = (\alpha + 1)\,\gamma + 1.$$

Die Berechnung der resultierenden Querkraft und des Biegemoments in einem beliebigen Querschnitt liefert:

$$Q = h \int_{-1}^{1} \sigma\, dz = B^{-\gamma}\, s^{\alpha\gamma}\, \dot{\varkappa}^{(k-1)}\, h^k\, k^{-1}\, \varphi,$$

$$M = h^2 \int_{-1}^{1} \sigma\, z\, dz = B^{-\gamma}\, s^{\alpha\gamma}\, \dot{\varkappa}^{(k-1)}\, h^{(k+1)}\, k^{-1} (k+1)^{-1} (y\,\varphi - k\,\psi).$$

Zur Abkürzung werden hier die Funktionen φ und ψ eingeführt, die folgendermaßen bestimmbar sind:

$$\varphi = (y+1)^k - (y-1)^k, \qquad \psi = (y+1)^k + (y-1)^k. \quad (3)$$

Für die Punkte A_1 und A_2 (vgl. Bild 1) werden die Gleichgewichtsbedingungen aufgestellt und gleichfalls berücksichtigt, daß das Biegemoment an der Übergangsstelle A_0 Null wird. Die Koordinaten des Punktes A_0 werden durch x_0, y_0 bezeichnet. Die Gleichgewichtsbedingungen für die Punkte A_1 und A_2 lauten:

$$q\, a_i = B^{-\gamma}\, h^k\, s_i^{\alpha\gamma}\, k^{-1}\, \dot{\varkappa}_i^{(k-1)}\, \varphi_i$$
$$0{,}5\, q\,(a_i^2 - x_0^2 - y_0^2) = \frac{h^{(k+1)}\, s_i^{\alpha\gamma}\, \dot{\varkappa}_i^{(k-1)}}{k\,(k+1)\, B^{\gamma}} (k\,\psi_i - y_i\,\varphi_i). \quad (4)$$

Der Index i, der hier und im folgenden die Werte 1 und 2 annimmt, weist darauf hin, daß die entsprechende Funktion für den Punkt A_1 oder A_2 gilt.

Das System (4) kann auf die Form

$$\varkappa_i = \left(\frac{a_i\, q\, k\, B^{\gamma}}{h^k\, s_i^{\alpha\gamma}\, \varphi_i}\right)^{\left(\frac{1}{k-1}\right)},$$
$$\frac{(k+1)(a_i^2 - x_0^2 - y_0^2)}{2\, h\, a_i} = k\,\frac{\psi_i}{\varphi_i} - y_i \quad (5)$$

gebracht werden.

Untersuchen wir den rechten Teil der zweiten der Gleichungen (5) genauer. Unter Benutzung von (3) kann man zeigen, daß für $y_i \gg 1$ diese rechte Seite mit einer Genauigkeit bis auf y_i^{-2} durch folgenden Ausdruck approximiert werden kann:

$$k\,\frac{\psi_i}{\varphi_i} - y_i \approx \frac{(k^2 - 1)}{3\, y_i}. \quad (6)$$

Selbst für $y_i = 1$ ist die Abweichung noch völlig unbedeutend ($(k^2 - 1)/3$ im Vergleich zum genauen Wert $(k - 1)$) außerdem ändert sich k selbst in den Grenzen $1 \div 2$, vgl. [3]). Folglich läßt sich die Beziehung (6) praktisch bis zu $y_i = 1$ benutzen. Andererseits gilt für $y_i \ll 1$

$$(k\,\psi_i/\varphi_i) - y_i \approx 1/y_i. \quad (7)$$

Die Ausdrücke (6) und (7) sind hinreichend einfach, um mit ihrer Hilfe φ_1 und φ_2 auszudrücken und, nach Einsetzen in (5), ein System von zwei gewöhnlichen nichtlinearen Differentialgleichungen zu erhalten, die sich nach den Ableitungen auflösen lassen. In diesem Fall hat das Gleichungssystem ein ziemlich einfaches Aussehen und läßt sich auf einem Digitalrechner bei Benutzung von Standardprogrammen integrieren.

Für den Fall einer kleinen Anfangsovalität des Ringes soll nun die Zeit t_* abgeschätzt werden, die vergeht, bis der kleinste Durchmesser Null wird. Aus [1] folgt, daß ein solcher Ring die kleine Ovalität im Verlaufe von ungefähr $0{,}9\, t_*$ beibehält. Einer kleinen Ovalität des Ringes entspricht der Fall $y_i \gg 1$. Aus (5) und (6) erhält man dann

$$1/y_i \approx 3\,(a_i^2 - x_0^2 - y_0^2)/(2\,(k-1)\, h\, a_i).$$

Für die Funktion φ_i gilt die Näherung

$$\varphi_i = 2\, k\, y_i^{(k-1)} \approx 2\, k \left[\frac{2\,(k-1)\, h\, a_i}{3\,(a_i^2 - x_0^2 - y_0^2)}\right]^{(k-1)}. \quad (8)$$

Durch Einsetzen von (8) in (5) erhalten wir schließlich:

$$\dot{\varkappa} = \left(\frac{1}{R_i}\right)^{\cdot} = \frac{3\,(a_i^2 - x_0^2 - y_0^2)}{2\,(k-1)\, h\, a_i} \left(\frac{q\, B^{\gamma}\, a_i}{2\, h^k\, s_i^{\alpha\gamma}}\right)^{\left(\frac{1}{k-1}\right)}. \quad (9)$$

Aus Bild 1 erkennt man, daß folgende Beziehungen gelten:

$$a_1 = R_1 + (R_2 - R_1) \sin\beta, \qquad a_2 = R_2 - (R_2 - R_1)\cos\beta,$$
$$x_0 = R_2 \sin\beta, \qquad y_0 = R_1 \cos\beta. \quad (10)$$

Es soll die Bedingung $\beta = \pi/4$ angenommen werden. Mit Einführen des mittleren Radius R_0 erhalten wir also

$$R_1 + R_2 = 2\, R_0, \quad x_0 = R_2/\sqrt{2}, \quad y_0 = R_1/\sqrt{2}. \quad (11)$$

Weiterhin soll ein Parameter Δ, der die Ovalität kennzeichnet, eingeführt werden:

$$a_1 = R_0\,(1 + \Delta), \qquad a_2 = R_0\,(1 - \Delta). \quad (12)$$

Der Ovalitätsparameter Δ_0 des elastischen Ringes hängt mit dem Parameter Δ_{00} des unbelasteten Ringes in folgender Weise zusammen:

$$\Delta_0 = \Delta_{00}\, q_E/(q_E - q), \qquad q_E = 2\, E\, h^3/R_0^3.$$

Dabei ist q_E der kritische Druck nach Euler [4] und E der Elastizitätsmodul.

Aus den Gleichungen (10) bis (12) erhält man

$$R_1 = R_0\,(1 - \Delta_1), \quad R_2 = R_0\,(1 + \Delta_1), \quad \Delta_1 = (1 + \sqrt{2})\,\Delta$$

und die Beziehung (9) wird auf die Form

$$\dot{\varkappa}_1 = \frac{3\, R_0\, \Delta}{(k-1)\, h\,(1+\Delta)} \left[\frac{q\, B^{\gamma}\, R_0\,(1+\Delta)}{2\, h^k\, s_1^{\alpha\gamma}}\right]^{\left(\frac{1}{k-1}\right)},$$

$$\dot{\varkappa}_2 = -\frac{3\, R_0\, \Delta}{h\,(k-1)(1-\Delta)} \left[\frac{q\, B^{\gamma}\, R_0\,(1-\Delta)}{2\, h^k\, s_2^{\alpha\gamma}}\right]^{\left(\frac{1}{k-1}\right)}$$

gebracht.

Daraus läßt sich näherungsweise die zeitliche Änderung von Δ ermitteln:

$$\frac{3\, R_0\, \Delta}{(k-1)} \left[\frac{q\, B^{\gamma}\, R_0}{2\, h^k}\right]^{\left(\frac{1}{k-1}\right)} = \frac{\dot{\Delta}_1^{\left(\frac{1}{\alpha+1}\right)}}{R_0},$$

$$\frac{\dot{\Delta}}{\Delta^{(\alpha+1)}} = D, \qquad D = (\sqrt{2} - 1)\, B \left(\frac{q\, R_0}{2\, h}\right)^n \left[\frac{3\, n\, R_0^2}{(\alpha+1)\, h^2}\right]^{\alpha+1}.$$

Die Integration ergibt

$$\Delta(t) = [\Delta_0^{-\alpha} - \alpha\, D t]^{-1/\alpha}.$$

Zur Ermittlung der Zeit t_* verwenden wir die Bedingung $\Delta(t_*) = 1$ und berücksichtigen außerdem die Bedingung $\Delta_0^{-\alpha} \gg 1$. Dann ergibt sich schließlich die Abhängigkeit der Zeit t_* von den Kennwerten des Materials, den Schalenparametern und der Größe des aufgebrachten Drucks in folgender einfacher Form:

$$t_* = (\alpha\, D\, \Delta_0^{\alpha})^{-1} \,. \tag{13}$$

Für den Fall des stationären Kriechens ($\alpha = 0$) ergibt die Beziehung (13) beim Grenzübergang $\alpha \to 0$ den Ausdruck:

$$t_* = \frac{1}{D} \ln \frac{1}{\Delta_0} \,.$$

Literatur

1 Ванько, В. И., Шестериков, С. А., Сплющивание кольца в условиях ползучести. МТТ, **5**, 127—130 (1966).

2 Локощенко, А. М., Шестериков, С. А., Релаксация труб и выпучивание стержней из вязко-пластического материала. ПМТФ, **4**, 154—159 (1966).

3 Шестериков, С. А., Об одном условии для законов ползучести. Изв. АН СССР, ОТН, Механи машиностр., **1**, 9 (1959).

4 Тимошенко, С. П., Устойчивость упругих систем. М., 1946.

Eingereicht am 23. 5. 1973

Anschrift: Локощенко, А. М., Москва В-234, МГУ, Институт Механики, отдел пластичности, СССР

ZAMM 54, 205 (1974)

H. E. FETTIS

An Alternative to Sylvester's Method

The most frequently used method of determining the consistency of two algebraic equations

$$a_0 + a_1\, x + a_2\, x^2 + \cdots + a_n\, x^n = 0 \,, \tag{1}$$
$$b_0 + b_1\, x + b_2\, x^2 + \cdots + b_m\, x^m = 0$$

is SYLVESTER's method [1], [2], [3]. This requires the formation of an $(m + n) \times (m + n)$ determinant which must vanish if the system (1) is consistent. If the a_i and b_i are functions of a second variable (say "y"), then this determinant set equal to zero is the resultant of (1).

For large values of m and n, the evaluation of the above mentioned determinant may become laborious, despite the fact that many elements are zero. The following procedure, requiring at each step only determinants of second order is suggested an an alternative.

Case I: The equations are of the same degree.

We first eliminate the constant terms between (1) and, after removing the factor x, obtain an equation of degree $(n - 1)$:

$$a_0^{(1)} + a_1^{(1)}\, x + a_2^{(1)}\, x^2 + \cdots + a_{n-1}^{(1)}\, x^{n-1} = 0 \tag{2}$$

where

$$a_0^{(1)} = a_0\, b_1 - b_0\, a_1 \,, \qquad a_1^{(1)} = a_0\, b_2 - b_0\, a_2 \,, \quad \text{etc.} \tag{3}$$

Next, we eliminate the term of highest degree between Eqs. (1):

$$b_0^{(1)} + b_1^{(1)}\, x + b_2^{(1)}\, x^2 + \cdots + b_{n-1}^{(1)}\, x^{n-1} = 0 \tag{2a}$$

where

$$b_0^{(1)} = a_0\, b_n - b_0\, a_n \,, \qquad b_1^{(1)} = a_1\, b_n - b_1\, a_n \,, \quad \text{etc.} \tag{3a}$$

Eqs. (2) and (2a) new constitute a system similar to (1) but of one lower degree. Repeating the process, we ultimately obtain two linear equations

$$a_0^{(n-1)} + a_1^{(n-1)}\, x = 0 \,, \qquad b_0^{(n-1)} + b_1^{(n-1)}\, x = 0 \,, \tag{4}$$

for which the solution is, obviously

$$a_0^{(n-1)}\, b_1^{(n-1)} - b_0^{(n-1)}\, a_1^{(n-1)} = 0 \,. \tag{5}$$

This method gives the final result with much less labor than would be required by the SYLVESTER determinant method, and in addition can be more readily adapted to programming for an automatic computer.

Case II: The equations are of different degree.

In this case, elimination of the constant term can be successively performed until the equations are of the same degree, whene the method of Case I is applicable.

References

1 TURNBULL, H. W., Theory of Equations, Oliver & Boyd, London 1946.

2 LOVITT, WM. V., Elementary Theory of Equations, Prentice-Hall, Englewood Cliffs, N. J. 1939.

3 KOWALEWSKI, G., Einführung in die Determinantentheorie, W. de Gruyter, Berlin 1942/Chelsea, New York 1948.

Eingereicht am 27. 6. 1973

Anschrift: HENRY E. FETTIS, 1885 California, Apt. 62 Mt. View, Ca. 94041, USA

ZAMM 54, 205 – 208 (1974)

R. N. KAPOOR/H. H. E. LEIPHOLZ

On Mass Distribution, Rotary Inertia and External Damping of a Viscoelastic Polygenic System

1. Introduction

In elastomechanical systems — possessing both the monogenic and polygenic forces as well as distributed and concentrated mass elements — some interesting results are due to PFLÜGER [1], DEINEKO and LEONOV [2], HAUGER [3] and LEIPHOLZ and LINDNER [4]. Monogenic forces — unlike polygenic ones — are derivable from scalar potentials. It has been shown that the critical load of these systems is greatly dependent on the mass distribution. Taking different cantilevered mathematical models with mass elements, the above investigators have shown that there exist critical mass ratios (in general, mass of element/mass of cantilever) for which the respective critical loads attain stationary values. In PFLÜGER's model — where the weight of the tip element is neglected — the flutter load first attains a relative minimum for a certain mass ratio and then an absolute maximum for infinite mass ratio. DEINEKO and LEONOV, taking two weightless mass elements, and varying their mass ratio arrive at results similar to PFLÜGER. In HAUGER's model — where the weight of the mass element is considered — an optimality condition is obtained i.e. for a certain mass ratio, the flutter load attains its maximum. LEIPHOLZ and LINDNER obtain similar optimality conditions. However, in dealing with these models, the external and internal damping as well as the rotary inertia of cantilever and mass elements have not been considered. The role of damping in polygenic systems can, at times, be paradoxical and should be considered. Similarly the influence of rotary inertia — especially when the mass ratios, as observed by these investigators, are high — cannot also be neglected. The work of HAUGER [3] and LEIPHOLZ and LINDNER [4] was already an improvement over that of PFLÜGER [1] and DEINEKO and LEONOV [2], in that, they took into account the inevitable influence of the weight of mass elements. This paper, falling into the third category, removes some more discrepancies between the mathematical model and the actual system — taking into account the influence of rotary inertia, and the material as well as external damping forces. The inspiration to write this paper stemmed from the principle author's doctoral work where, in course of investigating an altogether different class of viscoelastic polygenic systems, it was sensed that inclusion of certain additional realistic factors could lead not only to elimination of the aforementioned optimality characteristic but, more important, a substantial reduction in the value of incipient flutter loads and frequencies. The primary object of this paper, therefore, is to re-examine the HAUGER system

(Fig. 1)[1]) by taking into consideration: i) rotary inertia of the cantilever and tip element, ii) material damping of the cantilever in accordance with the improved HIGUCHI model [5], and iii) external damping of the cantilever as well as tip element. An exact analysis of the system differential equation leads to these significant results: a) mere inclusion of rotary inertia suppresses the critical loads and flutter frequencies — retaining the optimality characteristic, b) inclusion of material and external damping — in addition to rotary inertia — results in *substantial* destabilization of the system for *any* mass ratio, accompanied by further suppression of flutter frequencies. In addition, the optimality

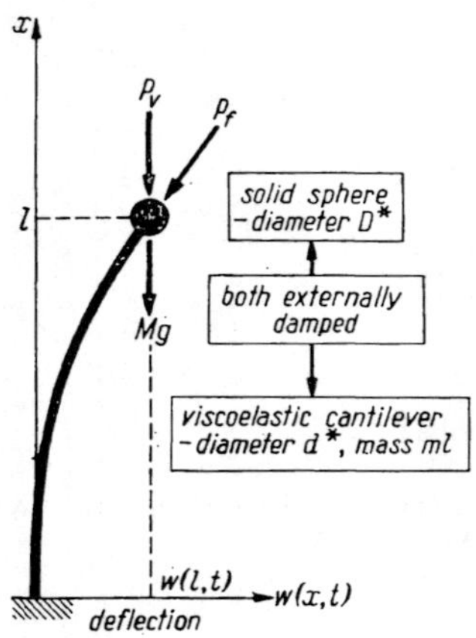

Fig. 1. The polygenic system

characteristic, as observed by previous investigators, *vanishes* completely. These results attain added significance in view of the fact that the destabilization observed here — unlike that due to fictitious infinitesimal damping — is for realistic damping values as observed in experiments [5].

2. The system and secular equations

For systems possessing the monogenic and polygenic forces, HAMILTON's equation is given by

$$\bar{\delta} \int_{t_1}^{t_2} \left(KE - SE + \sum_i \boldsymbol{F}^i_m \cdot \boldsymbol{D}^i_m\right) dt + \int_{t_1}^{t_2} \left(\sum_i \boldsymbol{F}^i_p \cdot \bar{\delta}\, \boldsymbol{D}^i_p\right) dt = 0 \tag{1}$$

where $\bar{\delta}$ is the variational notation, t the time, KE and SE the kinetic and strain energies, $\boldsymbol{F}^i_m$ and $\boldsymbol{F}^i_p$ the monogenic and polygenic forces and $\boldsymbol{D}^i_m$ and $\boldsymbol{D}^i_p$ the respective displacements. The various energy terms in Eq. 1 — pertaining to the modified HAUGER system, Fig. 1, are

$$KE = 0.5 \int_0^l (W_t^2 + k_1^2\, W_{tx}^2)\, dx + 0.5\, M\, (W_t^2 + k_2^2\, W_{tx}^2)_{x=l}$$

$$SE = 0.5\, E\, I \int_0^l W_{xx}^2\, dx$$

$$\sum_i \boldsymbol{F}^i_m \cdot \boldsymbol{D}^i_m = 0.5\, (P_v + P_f + M\, g) \int_0^l W_x^2\, dx \tag{2}$$

$$\sum_i \boldsymbol{F}^i_p \cdot \bar{\delta} \boldsymbol{D}^i_p$$

$$= - \int_0^l [\beta\, d^* W_t\, \bar{\delta} W + (\gamma_1\, W_{txx} + \gamma_2\, W_{ttxx})\, \bar{\delta} W_{xx}]\, dx - $$

$$- [(P_f\, W_x + 0.25\, \Pi\, D^{*2}\, \beta\, W_t)\, \bar{\delta} W]_{x=l}$$

where x is the coordinate along rod axis, μ the linear mass density, l the rod length, E the YOUNG's modulus, W the displacement, k_1 the radius of gyration of the bar cross section about the neutral axis, I the corresponding second moment, M the mass of the solid sphere, k_2 its radius of gyration about a diameter, g the acceleration due to gravity, P_v and P_f the vertical and follower tip forces, β the external viscous damping coefficient, γ_1 and γ_2 the coefficients of HIGUCHI material damping, d^* and D^* the rod and sphere diameters, and subscripts x and t of W the variables with respect to which partial derivatives are taken. The external damping forces are assumed proportional to velocity W_t as well as the frontal projected areas of rod and sphere. An increase of M results in a sphere of larger diameter, and a corresponding increase in the external resistance force on the sphere.

Substituting Eq. 2 in Eq. 1, and employing the dimensionless parameters

$$\eta = x\, l^{-1}, \quad \tau = (E\, I)^{1/2}\, (\mu\, l^4)^{-1/2}\, t, \quad R = \mu\, g\, l^3 (E\, I)^{-1},$$
$$\alpha = \gamma_1^{-1}\, \gamma_2 (E\, I)^{1/2}\, (\mu\, l^4)^{-1/2}, \quad \chi_1 = k_1^2\, l^{-2}, \quad \chi_2 = k_2^2\, l^{-2},$$
$$\chi_3 = 0.25\, \Pi\, D^{*2} (d^* l)^{-1}, \quad \varepsilon = M (\mu\, l)^{-1}, \quad M_s = \varepsilon\, R, \tag{3}$$
$$P_m = P_v l^2 (E\, I)^{-1}, \quad P_p = P_f\, l^2 (E\, I)^{-1}, \quad \delta_{n_1} = (\mu\, l^4\, E\, I)^{-1/2}\, \gamma_1,$$
$$\delta_{n_2} = \alpha\, \delta_{n_1}, \quad \delta_b = d^* l^4 (\mu\, l^4\, E\, I)^{-1/2}\, \beta$$

leads to the nonselfadjoint boundary value problem given by

$$W_{\eta\eta\eta\eta} + W_{\tau\tau} + \delta_{n_1} W_{\tau\eta\eta\eta\eta} + \alpha\, \delta_{n_1} W_{\tau\tau\eta\eta\eta\eta} + \delta_b W_\tau - $$
$$- \chi_1\, W_{\tau\tau\eta\eta} + (P_m + P_p + M_s) W_{\eta\eta} = 0 \tag{4}$$

and

$$(W = W_\eta)_{\eta=0} = 0,$$
$$(W_{\eta\eta} + \delta_{n_1} W_{\tau\eta\eta} + \alpha\, \delta_{n_1} W_{\tau\tau\eta\eta} + \varepsilon\, \chi_2\, W_{\tau\tau\eta})_{\eta=1} = 0,$$
$$[W_{\eta\eta\eta} + \delta_{n_1} W_{\tau\eta\eta\eta} + \alpha\, \delta_{n_1} W_{\tau\tau\eta\eta\eta} - \chi_1\, W_{\tau\tau\eta} - \chi_3\, \delta_b W_\tau - $$
$$- \varepsilon\, W_{\tau\tau} + (P_m + M_s)\, W_\eta]_{\eta=1} = 0. \tag{5}$$

Seeking a standing wave solution of the problem, in the form

$$W(\eta, \tau) = e^{\Omega \tau}\, \Phi(\eta) \tag{6}$$

where $\Omega = \Omega_R + j\, \Omega_I$ is the complex nondimensional frequency, and employing the additional dimensionless parameters

$$K_1 = 1 + \Omega\, (1 + \alpha\, \Omega)\, \delta_{n_1},$$
$$K_2 = (P_m + P_p + M_s - \chi_1\, \Omega^2)\, K_1^{-1},$$
$$K_3 = \Omega\, (\delta_b + \Omega)\, K_1^{-1}, \qquad K_4 = \varepsilon\, \chi_2\, \Omega^2\, K_1^{-1}, \tag{7}$$
$$K_5 = (P_m + M_s - \chi_1\, \Omega^2)\, K_1^{-1},$$
$$K_6 = -\, \Omega\, (\varepsilon\, \Omega + \chi_3\, \delta_b)\, K_1^{-1}$$

leads to the following simplified set of fourth order ordinary differential equations and boundary conditions

$$\Phi_{\eta\eta\eta\eta} + K_2\, \Phi_{\eta\eta} + K_3\, \Phi = 0, \tag{8}$$

$$\Phi(0) = \Phi_\eta(0) = 0,$$
$$\Phi_{\eta\eta}(1) + K_4\, \Phi_\eta(1) = 0, \tag{9}$$
$$\Phi_{\eta\eta\eta}(1) + K_5\, \Phi_\eta(1) + K_6\, \Phi(1) = 0.$$

Equation 8 admits the exact solution

$$\Phi(\eta) = \sum_{r=1}^{4} A_r\, e^{\lambda_r \eta}, \tag{10}$$

where

$$\lambda_{1,3} = \{0.5\, [-\, K_2 \pm (K_2^2 - 4\, K_3)^{1/2}]\}^{1/2}, \tag{11}$$
$$\lambda_2 = -\, \lambda_1, \qquad \lambda_4 = -\, \lambda_3.$$

By, in addition, satisfying the boundary conditions, Eq. 9, and seeking nontrivial solutions in A_r, results in the following complex secular determinant

$$\Delta_R + j\, \Delta_I = \begin{vmatrix} 1 & 1 & 1 & 1 \\ \lambda_1 & \lambda_2 & \lambda_3 & \lambda_4 \\ \Delta_1 & \Delta_2 & \Delta_3 & \Delta_4 \\ \Delta_1^* & \Delta_2^* & \Delta_3^* & \Delta_4^* \end{vmatrix} = 0, \tag{12}$$

where

$$\Delta_r = (K_4 + \lambda_r)\, \lambda_r\, e^{\lambda_r},$$
$$\Delta_r^* = (\lambda_r^3 + K_5\, \lambda_r + K_6)\, e^{\lambda_r}, \qquad (r = 1, 2, 3, 4). \tag{13}$$

[1]) It is assumed that P_v, P_f and Mg act simultaneously on the structure. Momentary absence of any of the above may alter the stability conditions appreciably [3], [4].

If ϱ_b and Δ_M denote the volumetric weight density of bar material and the density ratio of sphere and bar materials, respectively $-\chi_1$, χ_2 and χ_3, Eq. 3, are given by

$$\chi_1 = \varrho_b l(E R)^{-1}, \qquad \chi_2 = 0.1 (24 \varepsilon \chi_1)^{2/3} \Delta_M^{-2/3}, \qquad \chi_3 = \Pi(3 \varepsilon \chi_1^{1/4})^{2/3} (8 \Delta_M)^{-2/3}. \tag{14}$$

3. The stability boundary

We study the behaviour of eigenvalues P_p and Ω when, in Eq. 12, Δ_R and Δ_I vanish simultaneously. Inequality $\Omega_R < 0$ implies asymptotic stability and $\Omega_R > 0$ instability. Condition $\Omega_R = 0$ respresents dissimilar phases for the damped and undamped systems. For the former it implies *the instability boundary* and the latter *either the stable domain or the instability boundary*. F_{cr}, the critical value of end follower load P_p, is obtained from Eq. 12 by increasing P_p and observing when $\Omega_R = 0^+$, for the first time. The imaginary component Ω_I governs the vibrational behaviour of the system. Conditions $\Omega_I \neq 0$ and $\Omega_I = 0$ imply oscillatory and nonoscillatory motions, respectively. If at $\Omega_R = 0^+$, $\Omega_I = 0$, the system fails by divergence. At $\Omega_R = 0^+$, condition $\Omega_I \neq 0$ implies incipient flutter.

4. The illustrative example and discussion of results

Taking a steel rod 100 inches long, $R = 0.01$, $P_m = 1$, and $\Delta_M = 10$, the numerical results obtained on IBM 360 computer of the University of Waterloo are given in Table 1 and plotted in Figs. 2 and 3. Values of $\delta_{n_1} = 0.005729$ and $\delta_{n_2} = 0.002616$ correspond to the experimental values given for steel by HIGUCHI [5]. The value $\delta_b = 0.001$ is within acceptable range for normal surfaces vibrating in still air.

Table 1
The dependence of critical eigenvalues on the mass ratio

Mass ratio	Critical load			Flutter frequency		
	Case 1	Case 2	Case 3	Case 1	Case 2	Case 3
0.0	18.295	18.275	8.501	10.80	10.75	4.68
0.4	15.070	15.043	7.496	7.40	7.50	3.14
0.5	15.050	15.013	7.249	7.00	7.14	2.73
0.6	15.065	15.019	7.001	6.81	7.00	2.35
1.0	15.263	15.250	7.103	6.11	6.01	2.12
5.0	16.589	16.251	6.962	4.29	4.21	1.03
9.0	17.052	16.453	6.927	3.73	3.40	0.79
10.0	17.127	16.463	6.918	3.65	3.17	0.73
11.0	17.191	16.458	6.909	3.56	3.01	0.70
15.0	17.382	16.371	6.880	3.34	2.62	0.59
20.0	17.523	16.279	6.850	3.09	2.30	0.52
30.0	17.671	15.983	6.802	2.82	1.82	0.40
40.0	17.717	15.689	6.753	2.62	1.49	0.36
45.0	17.720	15.551	6.730	2.49	1.37	0.31
50.0	17.723	15.415	6.698	2.48	1.31	0.30
55.0	17.717	15.291	6.670	2.41	1.21	0.29
60.0	17.703	15.159	6.650	2.37	1.14	0.28
70.0	17.666	14.919	6.600	2.27	1.02	0.26

Case 1: Undamped without rotary inertia
Case 2: Undamped with rotary inertia
Case 3: Damped with rotary inertia

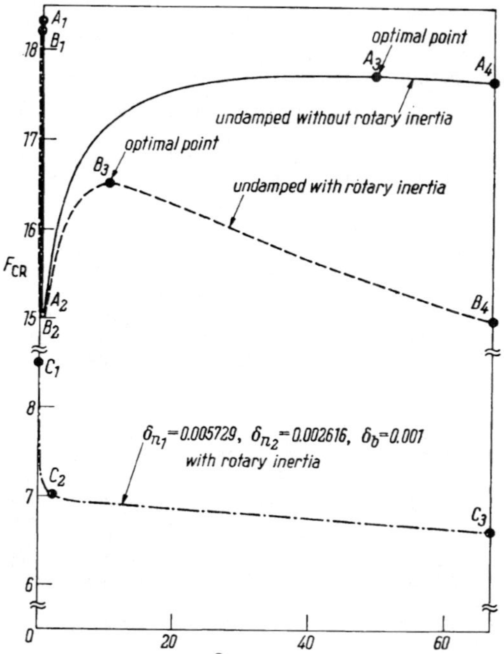

Fig. 2. Flutter loads vs. mass ratio

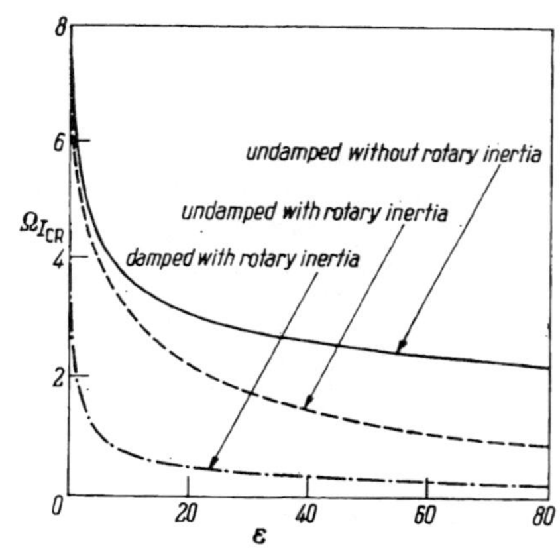

Fig. 3. Flutter frequency vs. mass ratio

The following observations are made from the figure:

I. Without rotary inertia and damping, upper curve $A_1A_2A_3A_4$, the flutter load F_{cr} is effected by the mass ratio ε, as follows:

a) initial increase of ε, portion A_1A_2, reduces F_{cr} sharply.

b) further increase of ε, portion A_2A_3, increases F_{cr} until the optimal value $F_{\text{cr}} = 17.724$ is attained at $\varepsilon = 49.9$.

c) further increase of ε, portion A_3A_4, lowers F_{cr}.

This is complete qualitative agreement with HAUGER's results [3] obtained by the approximate GALERKIN method.

II. With the inclusion of rotary inertia, curve $B_1B_2B_3B_4$ (under curve $A_1A_2A_3A_4$), F_{cr} is influenced by ε, as follows:

a) initial increase of ε, portion B_1B_2, has the same influence as in Ia) except that the critical loads are marginally reduced,

b) further increase of ε, portion B_2B_3, increases F_{cr} as in Ib) — though at a slower rate — and the reduced optimal flutter load $F_{\text{cr}} = 16.464$ is attained for a lower mass ratio, $\varepsilon = 10.2$.

c) further increase of ε, portion B_3B_4, lowers F_{cr} such that the gap between the first and second curves (destabilization due to rotary inertia) increases.

Hence, the overall influence of rotary inertia is to reduce the optimal values of flutter load and mass ratio — and increase the destabilization increase of ε.

III. With the inclusion of rotary inertia, as well as material and external damping, curve $C_1C_2C_3$, F_{cr} is influenced by ε, as follows:

a) for $\varepsilon = 0$, point C_1, the critical load $F_{\text{cr}} = 4.68$ is less than half the undamped value.

b) initial increase of ε, portion C_1C_2, reduces F_{cr} sharply as in Ia) and IIa).

c) further increase of ε, portion C_2C_3, does not — as in Ib) and IIb) — raise the flutter load which continues to decrease gradually. The optimality characteristic vanishes completely.

Hence, the overall influence of rotary inertia, as well as material and external damping, is to reduce the flutter loads substantially and eliminate the optimality characteristic observed in the previous two cases.

The flutter frequencies $\Omega_{I_{cr}}$ are influenced by ε in the following manner:

I. for the undamped system without rotary inertia, $\Omega_{I_{cr}}$ drops sharply for initial increase of ε. Further increase of ε tends to stabilize the drop.

II. with the introduction of rotary inertia, the flutter frequencies — as compared to I) — are not influenced appreciably for small ε. However, with increase of ε, the flutter frequencies reduce noticeably.

III. the combined influence of rotary inertia, as well as material and external damping, is to drastically reduce the flutter frequencies throughout — otherwise maintaining the qualitative pattern in I).

5. Conclusions

The authors, therefore, emphasize that stability analyses of flutter prone elastomechanical systems — possessing monogenic and polygenic forces as well as distributed and concentrated mass elements — are highly dubious unless the influence of rotary inertia and relevant damping forces are also accounted for. Paradoxically, these factors — as shown by an example in this paper — can reduce the critical loads and flutter frequencies, and therefore any results obtained without them may lead to erroneous and unsafe conclusions.

Acknowledgement

This research was supported by the National Research Council of Canada under grant number A 7297. The principle author gratefully acknowledges Ministry of Railways, Government of India to release him for the research work.

References

1 PFLÜGER, A., Zur Stabilität des tangential gedrückten Stabes, ZAMM **35**, p. 191 (1955).
2 DEINEKO, K. S., LEONOV, M. IA., A Dynamic Method for the Investigation of the Stability of a Compressed Bar, Prikl. Mat. i Mekhan. **19**, p. 738–744 (1955).
3 HAUGER, W., Bemerkungen zu dem Einfluß der Massenverteilung bei nichtkonservativen Stabilitätsproblemen elastischer Stäbe, Ing.-Arch. **35**, p. 283–291 (1967).
4 LEIPHOLZ, H., LINDNER, G., Über den Einfluß der Massenverteilung auf das nichtkonservative Knicken von Stäben, Ing.-Arch. **39**, p. 187–194 (1970).
5 HIGUCHI, S., On the Logarithmic Decrement of the Damped Oscillation of an Elastic Thin Beam, ZAMM **51**, p. 427–432 (1971).

Eingereicht am 27. 6. 1973

Anschriften: R. N. KAPOOR, Senior Mechanical Engineer, Indian Railways; Presently: Research Fellow, Solid Mechanics Division, Faculty of Engineering, University of Waterloo, Waterloo, Ontario, Canada, N2L 3Gl, Canada

H. H. E. LEIPHOLZ, Professor, Solid Mechanics Division, Faculty of Engineering, University of Waterloo Waterloo, Ontario, Canada, N2L 3Gl, Canada

BUCHBESPRECHUNGEN

Hrsg. R. Sauer und I. Szabó, Mathematische Hilfsmittel des Ingenieurs. Teil II. (Die Grundlehren d. math. Wissenschaften, B. 140). XX + 684 S. m. 148 Abb. Berlin/Heidelberg/New York 1969. Springer-Verlag. Preis geb. DM 136,—. Mathematische Hilfsmittel des Ingenieurs. Teil IV. (Die Grundlehren d. math. Wissenschaften, Band 142). XVIII + 596 S. m. 130 Abb. Berlin/Heidelberg/New York 1970. Springer-Verlag. Preis geb. DM 124,—.

Das Gesamtwerk „Mathematische Hilfsmittel des Ingenieurs", dessen Teile I und III in der ZAMM **49**, S. 568 (1969) besprochen worden sind, liegt mit den bald danach erschienenen Teilen II und IV vollständig vor. Was Absicht und Anlage des Werkes angehen, so kann auf die erste Besprechung verwiesen werden. Teil II umfaßt zwei, Teil IV drei Kapitel, deren Inhalt im folgenden kurz umrissen wird. Leider verbietet es der Platzmangel, auf die einzelnen Teile ausführlich einzugehen, und auch eine Rangordnung der behandelten Gegenstände nach ihrer Wichtigkeit für den wissenschaftlich-mathematisch arbeitenden Ingenieur ist schwer herzustellen, da sie stark vom Spezialgebiet des Betreffenden abhängt. Jedoch gewährt die Auswahl der Autoren von vornherein eine hervorragende fachliche Qualität.

W. Törnig behandelt in Teil D (292 S.) den umfassenden, für die Anwendungen in den Ingenieurwissenschaften so überaus wichtigen Komplex der *Anfangswertprobleme bei gewöhnlichen und partiellen Differentialgleichungen.* Der Leser findet nicht nur die Grundtatsachen, die auch in den für Ingenieure bestimmten mathematischen Grundvorlesungen berücksichtigt werden, sondern vor allem auch wichtige Hinweise zur praktischen, insbesondere numerischen Lösung der genannten Probleme. Von besonderer Wichtigkeit sind dabei natürlich die linearen Differentialgleichungen und Differentialgleichungssysteme. Bei den partiellen Differentialgleichungen ist es sachgemäß, daß die hyperbolischen und die parabolischen Differentialgleichungen unter Berücksichtigung ihrer Anwendungsgebiete mit den analytischen und numerischen Lösungsmethoden dargestellt werden.

Von nicht geringerer Bedeutung ist der von L. Collatz und R. Nicolovius verfaßte Teil E (377 S.), in dem es um die *Rand- und Eigenwertprobleme bei gewöhnlichen und partiellen Differentialgleichungen und bei Integralgleichungen* geht. Auch hier gehen Theorie und Praxis in vorbildlicher Weise Hand in Hand und wird auf die Anwendungen in wünschenswertem Maße eingegangen. Hervorgehoben seien die Kapitel über die Beziehungen zur Variationsrechnung, die in engem Zusammenhang mit numerischen Lösungsmethoden gebracht werden, über Differenzen- und Quadraturverfahren und über Iterationsverfahren.

Im vierten Band werden im Teil L (113 S.) von W. Hahn die für die Mechanik der Schwingungen so bedeutsamen Probleme der *Bewegungsstabilität bei Systemen mit endlich vielen Freiheitsgraden* behandelt, wobei auch auf die wichtige Ljapunowsche Theorie mit der gebührenden Ausführlichkeit eingegangen wird.

Im Teil M (134 S.) werden entsprechend der rasch zunehmenden Bedeutung, welche die *Wahrscheinlichkeitstheorie und Mathematische Statistik* auch in den Ingenieurwissenschaften gewinnen, die Grundtatsachen dieser Teilgebiete der Mathematik von D. Morgenstern und V. Mammitzsch in einer den Bedürfnissen der Ingenieure (und Physiker) angepaßten Auswahl (Allgemeines, Kombinatorik, Grundlegende Definitionen, Kombination von Ereignissen, Verteilungen von Zufallsgrößen, Abhängigkeitsmaße, Ungleichungen und Grenzwertsätze, Spezielle Verteilungen, Geordnete Stichprobe, Schätztheorie, Testtheorie, Lineare Modelle, Korrelations- und Regressionstheorie, Stichprobenverfahren, Stochastische Prozesse, Informationstheorie) dargeboten.

Der letzte, wesentlich umfangreichere Teil N (278 S.) bringt zunächst aus der Feder von W. Zander die grundlegenden *Sätze und Formeln der Mechanik* in einer stark formalisierten Darstellung, die die allgemeine Kontinuumstheorie (Grundelemente, Kinematik, Kräfte, Bilanzsätze, Materialgesetze) und Spezielle Theorien der Mechanik (Der starre Körper, Strömungsmechanik, Elastische Körper) umfaßt. K. Pöschl faßt in ähnlicher Weise die Grundlagen der *Elektrotechnik* zusammen und behandelt: Die Grundgleichungen des elektromagnetischen Feldes, Elektrische Potentialfelder, Stationäre magnetische Felder, Elektromagnetische Felder mit harmonischer Zeitabhängigkeit, Elektromagnetische Wellen und Schwingungen in Medien ohne freie Ladungen, Bewegungen geladener Teilchen in elektromagnetischen Feldern, Elektrisches Rauschen, Netzwerke und Übertragungssysteme, Signale und Signalerkennung.

Jeder Band ist für sich mit einem ausführlichen alphabetischen Sachregister ausgestattet, Band IV außerdem mit einem alle vier Bände umfassenden gemeinsamen Sachregister, das dem Benutzer des Gesamtwerkes besonders willkommen sein dürfte, da es die Orientierung in ihm und seine Handhabung als Nachschlagewerk ungemein erleichtert.

Dresden — H. Heinrich

A. Monjallon, Einführung in die moderne Mathematik. (Übers. a. d. Französischen). 163 S. m. 83 Bildern. Leipzig 1970. VEB Fachbuchverlag. Preis brosch. 14,80 M.

Mit der Herausgabe einer deutschen Übersetzung der 1963 bei Librairie Vuibert, Paris, in französischer Sprache erschienenen Originalfassung einer „Introduction aux mathématiques modernes" hat der VEB Fachbuchverlag Leipzig einem vermutlich großen Interessentenkreis einen guten Dienst erwiesen. Es wird über den im Vorwort zur deutschen Ausgabe besonders herausgestellten Zweck der Lehrerweiterbildung hinaus auch für Anfänger im Mathematikstudium von großem Nutzen sein, besonders dann, wenn diese in der Schule einen der modernen Auffassung noch nicht entsprechenden Mathematikunterricht erhalten haben. Der Inhalt beschränkt sich auf einfache Grundtatsachen. In den ersten drei Kapiteln werden die grundlegenden Begriffe, Relationen und Operationen der Mengenlehre erläutert. Ein viertes Kapitel beschäftigt sich mit Relationen, wobei besonders auf die Einführung in die mathematische Nomenklatur und ihre Handhabung eingegangen wird, ein fünftes mit dem Begriff der Funktion. Besonders wertvoll erscheinen dem Referenten das Kapitel 6 über die mathematische Sprache und das Kapitel 7, das mit „Ein wenig Axiomatik" überschrieben ist und dem Leser die Struktur mathematischer Beweise nahebringt und nützliche beweistechnische Hinweise gibt. Schließlich führt das achte Kapitel als Spezialfall einer mathematischen Struktur die kommutative Gruppe vor.

Das Schwergewicht liegt aber nicht auf dem mathematischen Gehalt, sondern auf der wohldurchdachten didaktischen Darstellung. Dem entspricht es auch, daß jedem Kapitel sorgfältig ausgewählte Übungsaufgaben beigefügt sind.

Gerade weil sich das Buch sicherlich bald über einen großen Interessentenkreis verbreitet haben wird, seien noch einige zusätzliche Bemerkungen gemacht:

Auf S. 41 (Mitte) ist die Formulierung: „$\mathcal{D}$ werde durch die Menge der Punkte im Innern und auf der Kurve $\mathcal{D}$ dargestellt" offensichtlich nicht korrekt, da hier $\mathcal{D}$ in zweierlei Bedeutung gebraucht wird. Auf S. 46, Zeile 8 v. o. muß es $A \cap [\bar{B} \cup C]$ statt $A \cap [B \cup C]$ heißen. Auf S. 61 wird gegen die sonst so stark vertretene Forderung der Strenge der mathematischen Sprache verstoßen, wenn aus einer Zeichnung „geschlossen" wird, daß .... Auf S. 67, Zeile 4 v. o., fehlt das Wörtchen „von" zwischen „Untermenge" und „$\mathcal{D} \times \mathcal{D}$". Auf S. 106, Zeile 10 v. o. sollte das Wort „Einige" durch Hinzufügung von „aber nicht alle" präzisiert werden. Diese Liste ist sicher nicht vollständig, zeigt aber, daß das Buch vor einer Neuauflage einer Durchsicht bedarf.

Noch ein Wort zu dem auf S. 124 eingeführten und weiterhin benutzten Begriff des „Modells". Man liest dort: „... so heißt das konkrete System ein *Modell* oder eine *Realisierung* des abstrakten Systems" und „*Jedes Theorem eines mathematischen Systems geht in jedem Modell des Systems in eine wahre Aussage über.*" Nach Auffassung des Referenten entspricht es mehr dem — wenigstens in den Anwendungsbereichen der Mathematik vertretenen — Standpunkt, wenn man umgekehrt das mathematische System als ein (mathematisches) Modell des konkreten Systems bezeichnete. Anwendung der Mathematik auf einen *vorgegebenen* „konkreten" Sachverhalt bedeutet dann das Auffinden eines mathematischen Modells, d. h. eines mathematischen Systems, dessen wahren Aussagen wahre (oder wenigstens näherungsweise gültige) Aussagen im konkreten System entsprechen.

Dresden H. HEINRICH

N. I. Mußchelischwili, Einige Grundaufgaben zu mathematischen Elastizitätstheorie. Carl Hanser Verlag München 1971; Übersetzung aus dem Russischen.

Der auf dem Gebiete der Angewandten Mathematik und Mechanik international bekannte Verfasser gibt in diesem Buche einen sehr umfassenden Überblick über die komplexen Lösungsverfahren für ebene lineare Elastizitätsprobleme der Scheiben, sowie für Probleme der Torsion und Querkraftbiegung prismatischer Stäbe. Jeweils werden die Randbedingungen und die daraus hervorgehenden Folgerungen eingehend erörtert und anhand von Anwendungsbeispielen demonstriert. Ein besonderes Kapitel ist den neuesten Forschungsergebnissen, u. a. der Rißtheorie gewidmet. Bei den heute stark verbreiteten numerischen Methoden ist ein Überblick über die analytischen Methoden besonders wichtig. Diverse Literaturbesprechungen erleichtern dem Leser das Verständnis.

München NEUBER

Hrsg. W. Dück und **M. Bliefernich,** Operationsforschung, 1. Mathematische Grundlagen, Methoden und Modelle. 440 S. m. 72 Abb. u. 34 Tabellen. Berlin 1971. VEB Deutscher Verlag der Wissenschaften. Preis je Bd. geb. 26,— M. Band 2: 491 S. m. 57 Abb. u. 65 Tab.

Die vorliegenden beiden Bände enthalten eine von 11 Autoren verfaßte Einführung in die mathematische Optimierung mit einer ausführlichen Darstellung der Grundlagen.

Der erste Band und der Anfang des zweiten umfassen die Kapitel: 1. Grundbegriffe der Logik und der Mengenlehre, 2. Differentialrechnung, 3. Integralrechnung, 4. Differentialgleichungen und Variationsrechnung, 5. Wahrscheinlichkeitsrechnung, 6. Lineare Algebra. Darin werden die Grundtatsachen aus der reellen Analysis und Gleichungslehre, sowie aus der Wahrscheinlichkeitstheorie zusammengestellt, soweit sie für die ökonomischen Anwendungen benötigt werden.

Die Darstellung wendet sich an mathematisch interessierte Ökonomen. Sie verzichtet dabei auf komplizierte Beweisführungen und legt das Hauptgewicht auf das Verständnis der mathematischen Sachverhalte und der daraus hergeleiteten Rechentechniken.

Der Rest des zweiten Bandes ist den Methoden der modernen Wirtschaftsmathematik gewidmet mit den Kapiteln: 7. Lineare Optimierung, 8. Stochastische lineare Optimierung, 9. Nichtlineare Optimierung, 10. Dynamische Optimierung, 11. Theorie der optimalen Prozesse, 12. Regelungstheorie, 13. Spieltheorie.

In diesen Kapiteln werden die neueren mathematischen Methoden entwickelt, die zur Beschreibung und Lösung ökonomischer Probleme, insbesondere der Optimierungsprobleme, benötigt werden. Die Verfahren sind anspruchsvoller als die Grundlagen des ersten Bandes, jedoch ist auch hier die Darstellung so gewählt, daß der mathematisch vorgebildete Wirtschaftswissenschaftler ihr gut folgen kann.

Zahlreiche Beispiele und Übungen (mit Lösungen) und ein Verzeichnis weiterführender Literatur ergänzen die gut lesbare Darstellung.

Bochum SOMMER

W. Tutschke, Grundlagen der reellen Analysis. I. Differentialrechnung: 205 S. m. 86 Abb. Preis brosch. 12,80 M. II. Integralrechnung: 161 S. m. 51 Abb. Preis brosch. 10,80 M. Berlin 1972. VEB Deutscher Verlag der Wissenschaften.

Das vorliegende Buch gibt eine präzise, systematische Darstellung der Differential- und Integralrechnung, wobei fast alle Betrachtungen sogleich im k-dimensionalen euklidischen Raum durchgeführt und vom reell-eindimensionalen Fall nur die Besonderheiten diskutiert werden. Auch die Verhältnisse im Komplexen werden parallel dazu mitbehandelt. Mit diesem ökonomischen Vorgehen wird der Autor seiner Zielstellung, vor allem theoretische Zusammenhänge herauszuarbeiten, voll gerecht. Da er sich hierbei eines außerordentlich knappen Stils bedient, ist das Buch in erster Linie zum Gebrauch neben einer entsprechenden Vorlesung bzw. für Leser mit gewissen Vorkenntnissen auf diesem Gebiet sehr nützlich. Die zahlreichen Beispiele und die 100 Übungsaufgaben mit Lösungen am Ende des entsprechenden Bandes sind zur Vertiefung der vermittelten Kenntnisse gut geeignet. (Das Buch enthält einige Druckfehler, z. B. in I, Seite 67, 10. Z. v. u.; II, Seite 30, 14. Z. v. o.; II, Seite 48, 13. Z. v. o.; II, Seite 90, 11. Z. v. u.)

In den ersten beiden Kapiteln werden die reellen und die komplexen Zahlen eingeführt sowie Punktmengen im Raum $\mathbb{R}^k$ untersucht.

Zwei weitere Kapitel befassen sich mit Grenzwerten von Funktionen und Eigenschaften stetiger Funktionen. Die folgenden beiden Kapitel sind der eigentlichen Differentialrechnung gewidmet. Dabei wird auch das Problem der Auflösbarkeit (eines Systems) von Gleichungen ausführlich behandelt. Das letzte Kapitel des ersten Bandes enthält die unendlichen Reihen — insbesondere Potenzreihen — im Komplexen. In diesem Zusammenhang werden auch die Exponentialfunktion und die trigonometrischen Funktionen definiert.

In den ersten beiden Kapiteln von Band II wird das RIEMANN-Integral reellwertiger Funktionen zunächst für k-dimensionale Intervalle definiert und sodann auf beliebige RIEMANN-meßbare Mengen erweitert. Eine sehr sorgfältige Behandlung erfährt die Berechnung k-dimensionaler Integrale. Das folgende Kapitel führt vom Begriff der Differentialform zur Integration über Mannigfaltigkeiten. Als Verallgemeinerung des Hauptsatzes der Differential- und Integralrechnung wird der Satz von STOKES-CARTAN bewiesen. Im letzten Kapitel wird das Maß von Mannigfaltigkeiten behandelt. Mit der Ableitung der Integralsätze der Vektoranalysis und deren physikalischer Interpretation schließt das empfehlenswerte Buch.

Dresden SCHIROTZEK

A. C. Pipkin, Lectures on Viscoelasticity Theory (Applied Mathematical Sciences, Vol. 7). IX + 180 S. m. 16 Fig. Berlin/Heidelberg/New York 1972. Springer-Verlag. Preis brosch. DM 20,80.

Die vom Verfasser an der Abteilung für Angewandte Mathematik der BROWN-Universität gehaltene einsemestrige Vorlesung über Viskoelastizität ist als Band der vom Springer-Verlag herausgegeben Reihe „Applied Mathematical Sciences" erschienen. Es stellt eine vorwiegend mathematisch orientierte Einführung in das Gebiet der Viskoelastizität dar, wobei mit Ausnahme der letzten beiden Kapitel die lineare Theorie betrachtet wird.

Die ersten drei Kapitel befassen sich mit den grundlegenden Beziehungen zwischen Spannungen und Verformungen sowie den Relaxations- und Kriechfunktionen, wobei ein Kapitel über FOURIER- und LAPLACE-Transformationen eingeschoben ist. Es folgen grundlegende Ausführungen über Torsionsschwingungen und Ausbreitung ebener Schubwellen im halbunendlichen Körper. Unter Benutzung der LAPLACE-Transformation werden Lösungen für die Grenzfälle des elastischen und des viskosen Materials und für den Zwischenbereich, dessen Kriech- und Relaxationsfunktion als durch ein Potenzgesetz darstellbar angenommen wird, abgeleitet.

Im fünften Kapitel geschieht der Übergang auf den räumlichen Fall bei Beschränkung auf quasi-statische Probleme und isotropes Material. Hier erfolgt eine gründliche Besprechung der elastisch-viskoelastischen Analogie (Korrespondenzprinzip). Das anschließende Kapitel geht von der Existenz einer Zeit-Temperatur-Äquivalenz aus und behandelt insbesondere die Auswirkungen, die sich aus dem Temperatureinfluß auf die mittlere Relaxationszeit ergeben.

Die letzten drei Kapitel befassen sich mit einigen Problemen großer Verformungen, wobei zunächst der Fall kleiner Verzerrungen bei großen Verschiebungen, dann nach einer

Klassifizierung viskoelastischer Strömungen auch der nichtlineare Fall der viskometrischen Strömung behandelt wird.

Gegenüber der mathematischen Formulierung als Grundlage der rechnerischen Behandlung von Problemen der Viskoelastizitätstheorie wird die Erörterung physikalischer Zusammenhänge auf das unbedingt Notwendige beschränkt. Das Buch richtet sich daher vorwiegend an Leser, die nicht ganz ohne Vorkenntnisse auf diesem Gebiete sind. Da es gleichzeitig nur eine beschränkte Stoffauswahl darstellen kann, wird vom Verfasser auf umfassendere Arbeiten verwiesen. Das Studium des Buches wird für den Leser vor allem dann besonders wertvoll, wenn er sich zu der vom Verfasser beabsichtigten Mitarbeit entschließt, die in der Ausführung der oft nur angedeuteten Ableitungen und der Beantwortung bzw. Lösung der im Text eingeschlossenen Fragen bzw. Probleme besteht.

Dresden BACKHAUS

R. v. Mises, Wahrscheinlichkeit, Statistik und Wahrheit (LEP Library of Exact Philosophy, Vol. 7). Vierte Aufl. 1 Porträt, XX + 293 S. Wien/New York 1972. Springer-Verlag. Preis geb. DM 69,—.

Fünfundvierzig Jahre nach dem Erscheinen dieses zu außerordentlicher Anerkennung gelangten Buches liegt nun die vierte im wesentlichen unveränderte Auflage vor, die von H. GEIRINGER durchgesehen wurde. Das Ideengut RICHARD VON MISES', das sich befruchtend auf die gesamte Entwicklung der Wahrscheinlichkeitstheorie und der Mathematischen Statistik — gerade in der Auseinandersetzung mit der fundamentalen Axiomatik A. N. KOLMOGOROFFS und dem R. A. FISHERschen Likelihood-Schätzprinzip — auswirkte, regt dabei immer wieder zum Nachdenken über die Grundlagen dieser mathematischen Disziplin an. Für jeden Wahrscheinlichkeitstheoretiker und Mathematischen Statistiker sollte deshalb das Erscheinen dieser neuen Auflage Anlaß sein, fragend an die Grundlagen seiner Wissenschaft heranzugehen.

Im übrigen sei auf die Besprechung der dritten Auflage durch Fr. A. WILLERS in dieser Zeitschrift Bd. 32 (1952), S. 31 verwiesen.

Dresden NOLLAU

M. Peschel/G. Wunsch, Methoden und Prinzipien der Systemtheorie. (Theoretische Grundlagen der Technischen Kybernetik). 1. Aufl. 132 S. m. Fig. Berlin 1972. VEB Verlag Technik. Preis brosch. 10,— M.

Dieses Buch macht bereits beim ersten Durchblättern einen völlig andersartigen Eindruck als die meisten anderen bekannten Bücher über Systemtheorie. Die Mittel der Beschreibung sind nicht so sehr FOURIER-, LAPLACE-, Z-, HILBERTtransformation und dergleichen, sondern allgemeinere Methoden aus der Funktionalanalysis und Algebra. Die zentralen Begriffe, von denen alle Überlegungen ausgehen, sind die Eingangs- und Ausgangssignale, der Zustand und die System-Operatoren, welche diese Größen einander zuordnen. Das erklärte Ziel der Autoren ist, auf diese Weise einer Vereinheitlichung der Systemtheorie näher zu kommen, welche kontinuierliche und diskrete, lineare und nichtlineare, deterministische und stochastische Systeme zugleich umfaßt.

Trier-Kaiserslautern RUPPRECHT

H. Triebel, Höhere Analysis. (Hochschulbücher für Mathematik, Band 76). 704 S. m. 42 Abb. Berlin 1972. VEB Deutscher Verlag der Wissenschaften. Preis geb. 60,— M.

Im Mittelpunkt des vorliegenden Buches steht die Behandlung linearer Differentialgleichungen auf der Grundlage der Spektraltheorie und der Theorie der Distributionen. Die benötigten funktionalanalytischen Hilfsmittel werden bereitgestellt; Ergebnisse aus der klassischen Theorie der (partiellen) Differentialgleichungen werden nicht verwendet. Dadurch ist das Buch trotz seiner hohen Anforderungen Studenten der Mathematik und der Physik ab 3. Studienjahr zugänglich, zumal die Darstellung außerordentlich klar und methodisch geschickt ist. (So wird beispielsweise vor der Darlegung eines komplizierteren Gedankenganges an dafür benötigte Begriffe und Sachverhalte, die bereits weiter vorn behandlet wurden, mit genauen Stellenangaben erinnert.) Von besonderem Interesse dürfte das Buch außer für Mathematiker, die sich mit der funktionalanalytischen Untersuchung von (partiellen) Differentialgleichungen befassen, auch für theoretische Physiker sein, denn der Autor geht nicht nur bei vielen Gelegenheiten auf physikalische Interpretationen ein, sondern er stellt auch in einem gesonderten Kapitel die mathematischen Grundlagen der Quantenmechanik dar. Sehr nützlich sind auch die in den Text eingefügten Hinweise auf weiterführende Literatur zu speziellen Problemen.

In den ersten beiden Kapiteln wird die Funktionalanalysis in BANACH- und HILBERTräumen in dem für das Weitere erforderlichen Umfang behandelt. Unter anderem wird die RIESZ-SCHAUDER-Theorie für kompakte Operatoren im HILBERTraum entwickelt und anschließend auf Integraloperatoren angewendet. Gleichzeitig enthalten diese Kapitel die entsprechenden Teile der Distributionentheorie. In dem folgenden Kapitel wird zunächst mit der Behandlung von Tensorprodukten und Faltungen die Darstellung der Distributionentheorie abgeschlossen. Auf der Grundlage der bisher bereitgestellten Resultate folgen sodann erste Untersuchungen über Anfangswertprobleme und Randwertprobleme für partielle Differentialgleichungen. Kapitel IV enthält die Spaktraltheorie selbstadjungierter (beschränkter und unbeschränkter) Operatoren sowie Ausführungen über Differentialgleichungen im HILBERTraum. Kapitel V ist einem sorgfältigen Studium gewöhnlicher Differentialoperatoren zweiter Ordnung mit Hilfe der Spektraltheorie und der Theorie der SOBOLEVschen Räume gewidmet. Die sog. speziellen Funktionen und ihre Eigenschaften werden daraus konsequent mit funktionalanalytischen Methoden gewonnen. Das folgende Kapitel befaßt sich mit elliptischen Differentialoperatoren und den zugehörigen Randwertproblemen. Unter Verwendung der SOBOLEVschen Einbettungssätze werden zuerst selbstadjungierte Differentialoperatoren und anschließend nicht-selbstadjungierte Operatoren als Störungen der entsprechenden selbstadjungierten Operatoren untersucht. Kapitel VII bringt, wie bereits erwähnt, eine exakte Darstellung der mathematischen Grundlagen der Quantenmechanik. Ausgangspunkt ist eine HILBERTraum-Axiomatik der Quantenmechanik, die ausführlich diskutiert wird. Besonders zu begrüßen ist dabei, daß der Autor sich nicht auf eine „rein mathematische“ Betrachtungsweise beschränkt, sondern auch die Frage der „Quantisierung“ und der physikalischen Interpretation eingehend erörtert. Zahlreiche quantenmechanische Probleme, wie z. B. die nichtrelativistische und relativistische Beschreibung des Wasserstoffatoms und die Periodizität des Systems der chemischen Elemente, werden mit spektraltheoretischen Methoden behandelt. Das letzte Kapitel enthält schließlich eine ausführliche Untersuchung von Anfangswertproblemen und Rand-Anfangswertproblemen für parabolische und hyperbolische Differentialgleichungen.

Drei kurze Anhänge (Maß- und Integrationstheorie, Γ-Funktion, Integralformeln von GAUSS und GREEN) beschließen dieses empfehlenswerte Buch, das die große Wirksamkeit funktionalanalytischer Methoden eindrucksvoll demonstriert.

Dresden SCHIROTZEK

H. Richter/V. Mammitzsch, Methode der kleinsten Quadrate. 144 S. m. Übungen u. Aufgaben. Stuttgart/Berlin/Köln/Mainz 1973. Verlag Berliner Union GmbH./Verlag W. Kohlhammer GmbH. Preis brosch. DM 24,—.

Im vorliegenden Buch wird die „Methode der kleinsten Quadrate“ von einem axiomatischen Standpunkt aus behandelt. Das von C. F. GAUSS begründete Schätzprinzip — nämlich mittels eines Minimalprinzips die Punktschätzungen „wahrer Parameter“ zu definieren — wird von den Verfassern als formal und somit unbefriedigend angesehen, ebenso wie die Motivation für lineare Punktschätzungen mittels des GAUSS-MARKOV-Theorems. Die Verfasser gehen vielmehr von folgenden drei Axiomen aus:

1. Besteht zwischen den Meßgrößen $\alpha_1, \ldots, \alpha_n$ kein funktionaler Zusammenhang und besitzt man für jede Größe α_i genau eine Messung a_i, so gilt a_i als bester Ersatz für das unbekannte α_i für $i = 1, \ldots, n$.
2. Sind die Zufallsvektoren $\boldsymbol{a}$ und $\boldsymbol{c}$ unabhängig und besteht zwischen den zufälligen Vektoren der Meßgrößen $\boldsymbol{\alpha}$ und $\boldsymbol{\gamma}$ kein funktionaler Zusammenhang, so hängt der beste Wert für das unbekannte $\boldsymbol{\alpha}$ nur von $\boldsymbol{a}$, nicht aber von $\boldsymbol{c}$ ab.
3. Besteht zwischen den Zufallsvektoren $\boldsymbol{a}$ und $\boldsymbol{b}$ die umkehrbare lineare Beziehung $\boldsymbol{b} = \boldsymbol{\varphi}(\boldsymbol{a})$, so gelten für den besten Ersatz $\boldsymbol{\alpha}^*$ reps. $\boldsymbol{\beta}^*$ von $\boldsymbol{\alpha} = E\,\boldsymbol{a}$ reps. $\boldsymbol{\beta} = E\,\boldsymbol{b}$ die Beziehungen $\boldsymbol{\alpha}^* = \boldsymbol{\varphi}^{-1}(\boldsymbol{\beta}^*)$ bzw. $\boldsymbol{\beta}^* = \boldsymbol{\varphi}(\boldsymbol{\alpha}^*)$.

Die Verfasser erhalten mittels dieser Axiome die bekannten Beziehungen und Aussagen für die Punktschätzungen der unbekannten Parameter und leiten einige wichtige Resultate zur Genauigkeit dieser Punktschätzungen im Falle einer Ausgleichsrechnung bei mehrfacher Beobachtung einer Größe bei indirekten Beobachtungen und bei bedingten Beobachtungen her.

Die gesamte Darstellung folgt einem neueren Trend in der Statistik der linearen Modelle, nämlich der konsequenten Verwendung des Matrizenkalküls. Die Verfasser kommen dabei dem Leser entgegen, indem sie eine Einführung in die Matrizenrechnung geben, soweit sie hier von Interesse ist. Eine größere Anzahl von Aufgaben (mit Lösungen) erweisen sich gewiß als wertvoll bei der Aneignung dieses Wissensgebietes. Alles in allem liegt mit diesem Buch wiederum ein interessanter Beitrag zur modernen Darstellung klassischer Methoden der Mathematischen Statistik vor.

Dresden NOLLAU

A. Spătaru, Theorie der Informationsübertragung. In deutscher Sprache herausgegeben von P. Fey. (Elektronisches Rechnen und Regeln, Sonderband 18). XXII + 696 S. m. 418 Abb. Berlin 1973. Akademie-Verlag. Preis geb. 78,— M.

Für das Gebiet der Nachrichten- oder Informationsübertragung beginnt sich eine allgemein gebräuchliche Darstellungsweise herauszukristallisieren, welche sowohl die älteren Teilgebiete der Modulation und der Beschreibung deterministischer Signale wie auch die neueren Teilgebiete der Beschreibung zufälliger Signale und Störungen sowie der Informationstheorie umfaßt. Das vorliegende Werk des rumänischen Autors stellt eine sehr beachtenswerte Wegmarke in dieser Entwicklung dar.

Nach einem einführenden Abschnitt über die Grundbegriffe der Wahrscheinlichkeitsrechnung, welche die mathematische Grundlage zur Beschreibung von Signalen, Störungen und Information darstellt, folgen zunächst drei Abschnitte über Signale. Deren erster befaßt sich mit deterministischen Signalen und behandelt u. a. die FOURIER- und LAPLACE-transformation, das Abtasttheorem und das analytische Signal. Deren zweiter beschreibt die Eigenschaften zufälliger Signale, darunter ihre Spektraldichte, Autokorrelationsfunktion und orthogonale Reihenentwicklung und deren dritter die Signaldarstellung im Euklidischen Raum und die Signalflußgraphen.

Auf diese ersten Abschnitte, die knapp 30% des Buchumfangs ausmachen, folgen drei weitere Abschnitte über Systeme, und zwar über lineare stationäre Systeme, welche in vielen Beispielen behandelt werden, über lineare zeitvariable Systeme, einschließlich der diskreten zeitvariablen Systeme und der hierfür wichtigen Z-Transformation und schließlich über nichtlineare Systeme. Auf einen relativ kurzen Abschnitt über Störungen folgt dann eine recht ausführliche Darstellung der Modulation, welche wieder etwa weitere 30% des Buchumfangs beansprucht. Dieser Teil besteht aus den Abschnitten über lineare Modulation, worunter die verschiedenen Versionen der Amplitudenmodulation eines Sinusträgers verstanden werden, über exponentielle Modulation, durch welchen Begriff die Frequenz- und Phasenmodulation eines Sinusträgers zusammengefaßt werden, und über Pulsmodulation. Der Darstellung der Modulation folgt eine Beschreibung verschiedener Multiplex-Systeme. Abgeschlossen wird das Buch mit zwei Abschnitten über die Information diskreter und kontinuierlicher Signale.

Das Werk wird seinem Anspruch, ein Lehrbuch zu sein, in hervorragender Weise gerecht. Der Text ist klar und leicht verständlich abgefaßt. Das Niveau entspricht dem der an deutschen Hochschulen üblichen Fachvorlesungen der Studienrichtung Nachrichtentechnik nach dem Vordiplom.

Trier-Kaiserslautern RUPPRECHT

Die besprochenen und angezeigten Bücher sind durch den Buchhandel zu beziehen.

Bestellungen sind an das zuständige Postamt oder eine Buchhandlung zu richten, Bestellungen aus dem Ausland an die Firma Buchexport, Volkseigener Außenhandelsbetrieb der DDR, 701 Leipzig, Leninstr. 72, oder direkt an den Akademie-Verlag, DDR-108 Berlin, Leipziger Str. 3—4

Herausgeber und Hauptschriftleiter: Prof. Dr.-Ing. habil. Dr. techn. h.c. H. Heinrich, 8027 Dresden, Friedrich-Hegel-Str. 31. Verlag: Akademie-Verlag, DDR-108 Berlin, Leipziger Straße 3—4; Fernsprecher: 220441, Telex-Nr. 112 020, Postscheckkonto: Berlin 35021. Bestellnummer dieses Heftes: 1009/54/3. Die Zeitschrift für Angewandte Mathematik und Mechanik erscheint jährlich mit 12 Heften, einschließlich Tagungsheft. Bezugspreis je Heft 22,— M (Sonderpreis für die DDR 18,— M); Bezugspreis eines Bandes 264,— M zuzüglich Versandkosten (Sonderpreis für die DDR 216,— M). Veröffentlicht unter Lizenznummer 1282 des Presseamtes beim Vorsitzenden des Ministerrates der Deutschen Demokratischen Republik.
Gesamtherstellung: VEB Druckerei „Thomas Müntzer", 582 Bad Langensalza.
Printed in the German Democratic Republic.

FSC
www.fsc.org
MIX
Papier | Fördert gute Waldnutzung
FSC® C083411

Zeitfracht Medien GmbH
Ferdinand-Jühlke-Straße 7
99095 Erfurt, Deutschland
produktsicherheit@kolibri360.de